The Memory Jogger Plus+®

Featuring the Seven Management and Planning Tools

by Michael Brassard

Revised Edition

© **1996, 1989 GOAL/QPC**
13 Branch Street
Methuen, MA 01844
1-800-643-4316
978-685-3900
Fax: 978-685-6151
E-mail: service@goal.com
Web site: http://www.goalqpc.com

First Edition
15 14 13 12 11

ISBN 1-879364-83-2

Dedication

To my wife, Jaye, for her love, support, and <u>patience</u>
and to my two buddies, Nina and Paige,
for the diversions they've provided.

Acknowledgments

Special thanks to the following people for all of their suggestions, insights, and examples during the development of *The Memory Jogger Plus+®*.

Don Andrews, Dow Chemical
Jack Moran, Polaroid Corporation
Don Murphy, Hughes Aircraft
Phil Dargie, VELCRO USA
Michael Grimes, Lakeshore Technical College
Richard Zultner, Zultner & Associates
Jim Naughton, Expert Knowledge Systems
Karen Jamrog, GOAL/QPC Staff

To all of the course attendees over the last five years who have taught me so much.

To Bob King for his patience and support for my independent methods and trust in my judgment.

A real debt of gratitude to Diane Ritter for her invaluable insights and guidance on both the content and style of the book and for putting up with me. All of the graphics and layout are her products.

Cover Design by Tom Beaulieu of Absolute Design & Consulting, Boston, MA. 1996 Revised Edition was worked on by Michele Kierstead (design and layout) and Francine Oddo (editing and proofreading).

Contents

Author's Note

In the course of developing this book and using these tools for the last five years, the most often asked questions have been:

1) Are these tools really new?

2) Were they made in the United States or Japan?

3) Do line and staff managers really have time to use them on a regular basis?

Once again, some truisms are true. There is nothing truly new since virtually everything is a variation on a theme. Some of the tools (Tree Diagram, Matrix Diagram, and Activity Network Diagram) are no different from their American versions. Others [Affinity Diagram, Interrelationship Digraph, Prioritization Matrices, and Process Decision Program Chart (PDPC)] bear a strong resemblance to American techniques but have developed unique features that seem to really enhance their usefulness. Therefore, the answer to the second question about their country of origin is "yes." However, there are two things that are indeed new:

1) These tools have been combined into a cycle of activity that turns the output of one tool into the input of a related technique. This creates a continuous flow of analysis that really focuses any planning process.

2) The 7 MP Tools, which have been used in isolation by planning specialists in the past, are now available to the mainstream manager. This integrated approach can help any manager become a better planner and implementer. Up to this point these managers have been left to their own devices. Some have become effective planners; most have not.

Finally, there is the question of time. Managers who effectively use these tools are those individuals who look at planning itself as a time saver in terms of total implementation time. Focused planning up front can prevent expensive and time consuming "rework" when a poorly planned implementation fails. This does present somewhat of a dilemma: many of these managers who don't like to plan may feel this way because they haven't had a set of effective planning tools. Hopefully, the non-planners among us will give both the 7 MP Tools and planning a chance to work together. If they do, time will somehow be found to do both. Good luck!

Preface

In January of 1984, a small group of dedicated GOAL/QPC members formed the Statistical Resource Committee. Its task was to research, review, and redesign, if necessary, the written and video materials available at that time on Statistical Process Control (SPC). More specifically, it was to look at any and all SPC training materials available for the education of hourly workers, supervisors, and middle and top management.

After a number of meetings, the Committee agreed that there was a wide variety of materials available (some better than others) for the **academic** teaching of SPC, but nothing for the practitioner to use day in and day out. By *"practitioner,"* we mean the workers and managers who are standing by a machine or sitting behind a desk who are given the task of doing something to control the variation in that machine or to correct a paperwork system gone haywire.

Eureka! The birth of *The Memory Jogger*™! After about a year of development, GOAL/QPC published this small, yet jam-packed book. We must have hit a responsive chord because at the time of this writing, we have sold over one million of them in the United States and *The Memory Jogger* is now available in Spanish, French, German, and Italian.

So much for *The Memory Jogger*. Why create *The Memory Jogger Plus+*®? While we were developing *The Memory Jogger* in 1984, we were also becoming acquainted with what were called the "Seven New Tools for Quality Control." The existence of these "New Tools" was mentioned in passing by Jogi Arai, the Director of the Japan Productivity Center's Washington office. In an informal conversation following one of his speeches, he casually mentioned that the Japanese were using some techniques that enabled some companies to reduce their design cycle time by about half. In a study trip to Japan, Bob King, GOAL/QPC's Executive Director, discovered Japanese texts written on two critical areas: Quality Function Deployment (QFD) and the 7 New QC Tools. The *7 New Quality Tools for Managers & Staff* (1979) by Mizuno was the only source we could locate in our research. We translated it into a working text from which we could learn and develop our own skills in using these tools. Eventually, this same book was again translated, edited, and published as a fine text by Productivity Press, Cambridge, Massachusetts under the title *Management For Quality Improvement: The 7 New QC Tools* (1988).

Since the time of translation, we have evolved from using the term "7 New QC Tools" to the "Seven Management and Planning Tools." We made this change because the former seemed to suggest that these tools were replacing the 7 "Old Tools" (Pareto Chart, Cause and Effect Diagram, Flow Chart, etc.). The revised title seemed more descriptive and suggestive of a supplement, not a replacement, of the existing tools. Hopefully, this book will show that these two sets of tools are complementary, not competitive.

(While this new title may be descriptive, it is also cumbersome. Hereafter, the tools will be referred to as the **7 MP Tools**.)

Since 1984, GOAL/QPC has been teaching these tools to a number of U.S. companies while refining our own understanding of their construction and many uses. During this period GOAL/QPC has also been working hard to develop American (rather than Japanese) examples. We feel that it is time to share this experience and knowledge with those who must "make it happen." Therefore, *The Memory Jogger Plus+*® is designed to provide the same practical help in using the 7 Management and Planning Tools that *The Memory Jogger* did with the 7 Basic QC Tools.

How to Use *The Memory Jogger Plus+*®

The first thing that you'll notice about *The Memory Jogger Plus+*® is that it is considerably bigger than *The Memory Jogger*. This extra space is needed because the 7 MP Tools use words and phrases displayed in complex-looking charts instead of data points summarized and graphed in the Basic QC Tools like the histogram, scatter diagram, control chart, etc. The examples of the 7 MP Tools simply couldn't be reduced to *The Memory Jogger* size and still be legible. However, the real beauty of *The Memory Jogger* is its size. You can just slip it in your pocket or pocketbook and it's readily available for quick reference.

The Memory Jogger Plus+® is filled with the following useful information:

- The history of the 7 MP Tools and their relationship to the 7 Basic QC Tools.
- A complete case study that details the Frankel Corporation's use of the 7 MP Tools to tackle tough issues in both product-oriented and administrative areas. The case study shows illustrations of each step in the use of all of the 7 MP Tools.
- A variety of examples of each of the tools in their different possible formats.
- Helpful hints to maximize the usefulness of each tool.
- Selection criteria that will help you select the right tool for a wide variety of situations.

With this easy-to-use format you can acquaint yourself with each of the tools, see examples of its use, evaluate your own situation and problem-solving needs, and select the best available technique to guide you in your first attempts to use these tools.

We at GOAL/QPC hope that the 7 MP Tools and *The Memory Jogger Plus+®* will make the process of daily continuous improvement just a little bit easier. Good luck!

Michael Brassard
Methuen, MA
December 1989

— *Notes* —

Introduction

Seven Management and Planning Tools for Continuous Quality and Productivity Improvement

In 1950, Dr. Deming drew the following diagram on the blackboard during his first meeting with the Japanese Union of Scientists and Engineers (J.U.S.E.).

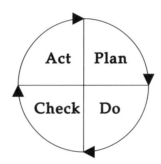

Dr. Deming called it the Shewhart Cycle. It has since become known as the Deming Cycle (or Plan-Do-Check-Act Cycle). It graphically describes the action steps we use every day to manage our lives and our businesses.

1. We **PLAN** what we want to accomplish over a period of time and what we are going to do to get there.

2. We **DO** something that furthers the goals and strategies developed in number one.

3. We **CHECK** the results of our actions to make sure there is a close fit between what we hoped to accomplish and what was actually achieved.

4. We **ACT** by making changes that are needed to more closely achieve the initial goals or by developing procedures to ensure continuance of those plans that were successful.

Problems in Implementing this Cycle

Even though this is a *natural* process, its effective application in many U.S. companies has been limited by several major factors.

a) Since the days of Frederick Taylor, the planning and evaluation functions have been separated from the *doers*. This was based on the principle that technical specialists (e.g., industrial engineers, Q.A. engineers) knew best how to plan to do a job while an unskilled work force knew only how to execute the plan. This eventually led to the strict departmentalization of job functions that is so prevalent today in the United States and other western countries.

b) Planning has often been relegated to the "seat of the pants" approach. It has often been seen as either too theoretical to be of practical use or too detailed and mundane. It has not been viewed as being where the action is. There is the common perception "that doers are recognized and rewarded while planners just plan."

c) There has been a lack of available tools that make the job of planning simple and timely.

Impact of the "Total Quality" Revolution

The "Total Quality" Revolution that is underway today is addressing these issues. The Deming Philosophy, Total Quality Control (TQC), and Total Quality Management (TQM) focus heavily on breaking down these organizational barriers to improvement. Based on these approaches, Taylorism in many U.S. companies is an endangered species.

The "Seven Management and Planning Tools" finally provide every manager with the tools needed to make planning an effective and satisfying process. They also break down Taylor-type barriers by giving more individuals the ability to contribute to the planning step.

History of the 7 MP Tools

Most of the 7 MP Tools are not new at all. Rather, most of them have their roots in post-World War II Operations Research work as well as in the work of leaders in Japanese TQC. This Japanese research effort was conducted by a committee of the Society for QC Technique Development. Between 1972 and 1979, this committee refined and tested individual tools and the overall cycle.[1] Finally, in 1979 the book *Seven New Quality Tools for Managers and Staff* was published. As mentioned earlier, GOAL/QPC discovered this book in 1983, had it translated, and began teaching the 7 MP Tools in 1984. The rest is history.

---

[1] Mizuno, Shigeru, *Management for Quality Improvement: The 7 New QC Tools*, Productivity Press, Cambridge, MA, 1988, pp. xi-xii.

What Are These Mysterious 7 MP Tools?

The 7 MP Tools are neither complex nor mysterious and include:

1. Affinity Diagram/KJ Method®*
2. Interrelationship Digraph (I.D.)
3. Tree Diagram
4. Prioritization Matrices
5. Matrix Diagram
6. Process Decision Program Chart (PDPC)
7. Activity Network Diagram

Tool Description

The 7 MP Tools are:

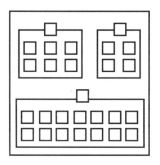

Affinity Diagram: This tool gathers large amounts of language data (ideas, opinions, issues, etc.) and organizes it into groupings based on the natural relationship between each item. It is largely a creative rather than a logical process.

---

* The name "KJ Method" is a registered trademark of the originator, Jiro Kawakita. To request copyright permission contact Kawakita Laboratory 1-3-20-801, Shimo Meguro, Tokyo 153, Telephone: 03-493-5725.

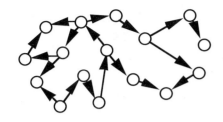

Interrelationship Digraph (I.D.): This tool takes complex, multi-variable problems or desired outcomes and explores and displays all of the interrelated factors involved. It graphically shows the logical (and often causal) relationships between factors.

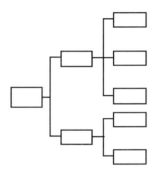

Tree Diagram: This tool systematically maps out in increasing detail the full range of paths and tasks that need to be accomplished in order to achieve a primary goal and every related subgoal. Graphically, it resembles an organization chart or family tree.

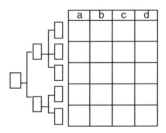

Prioritization Matrices[1]: These tools take tasks, issues, or possible actions and prioritize them based on known, weighted criteria. They utilize a combination of Tree and Matrix Diagram techniques, thus narrowing down options to those that are the most desirable or effective.

---

[1] This replaces Matrix Data Analysis in the original set of seven tools. A complete explanation of this change is found in Chapter 4.

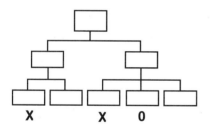

Matrix Diagram: This versatile tool shows the connection (or correlation) between each idea/issue in one group of items and each idea/issue in one or more other groups of items. At each intersecting point between a vertical set of items and horizontal set of items a relationship is indicated as being either present or absent. In its most common use the Matrix Diagram takes the necessary tasks (often from the Tree Diagram) and graphically displays their relationships with people, functions, or other tasks. This is frequently used to determine who has responsibility for the different parts of an implementation plan.

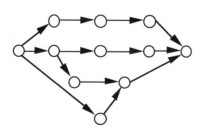

Process Decision Program Chart (PDPC): This tool maps out every conceivable event and contingency that could occur when moving from a problem statement to the possible solutions. This is used to plan each possible chain of events that needs to happen when the problem or goal is an unfamiliar one.

Activity Network Diagram[2]: This tool is used to plan the most appropriate schedule for any complex task and all of its related subtasks. It projects likely completion time and monitors all subtasks for adherence to the necessary schedule. This is used when the task at hand is a familiar one with subtasks that are of a known duration.

---

[2] This replaces the Arrow Diagram in the original set of seven tools. A complete explanation of this change is found in Chapter 7.

How Do the Tools Flow Together?

One thing that the Japanese do very well is to make techniques more powerful by combining them into a cycle of activity in which the output of one technique becomes an input to the next technique. For example, each of the Basic QC Tools (Check Sheet, Pareto Chart, etc.) can be used individually. However, the real power comes when a Check Sheet is converted into a Pareto Chart which in turn provides the focus for the Cause and Effect Diagram, etc. The same is true for the 7 MP Tools. Each of the seven techniques can be used alone very effectively. However, the full effect comes when they are used together to move from a chaotic situation to an implementable action plan for improvement.

The following is simply a "typical" flow to demonstrate how the tools usually flow. Be creative in your combinations and be disciplined in the use of each tool.

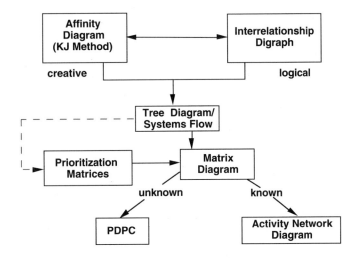

**7 Management & Planning Tools
Typical Flow**

The process can start at either the Affinity Diagram or the Interrelationship Digraph (I.D.). The Affinity is appropriate when there is a need to identify broad issue categories while the I.D. identifies base causes. Either can be used to narrow down the focus to one or more key issues for further planning. The Affinity does this through the consensus of the team. The I.D. naturally produces this by examining the pattern of arrows.

The Tree Diagram takes this identified issue and breaks it down into greater and greater implementation detail. The lowest level of the Tree (the most detailed) is either a listing of options (things that could be done) or a universal listing (all things that must be done). In the case of options, Prioritization Matrices can be used to compare all of the options based on known criteria. For universal listings, the various forms of the Matrix Diagram can be used (e.g., a Responsibility Matrix).

Once the Matrix has identified key items or responsibilities, the Process Decision Program Chart (PDPC) can be used to anticipate likely contingencies and reasonable countermeasures that must be built into any implementation plan. The PDPC is used when the implementation tasks are unfamiliar.

Also after completing the Matrix Diagram, a team can choose to use an Activity Network Diagram to map out familiar tasks with known paths and duration. A PDPC can also be converted into an Activity Network Diagram for scheduling and monitoring.

In the end, the planning path depends on the team's need. Control the process to fit the task; don't force fit the task into a neat, but inappropriate model.

Who Uses the 7 MP Tools?

The 7 MP Tools have proven useful to managers of virtually any level in a company. However, they have proven to be particularly helpful for middle to upper managers. They seem to be especially acceptable at these levels because they fill a void left by the Seven Basic QC Tools (Pareto Chart, Cause and Effect Diagram, etc.). These original tools appear to have several disadvantages in the eyes of many middle- to upper-level managers:

1. The simple, graphical techniques such as the Flow Chart, Check Sheet, Trend Chart, Pareto Chart, and Cause and Effect Diagram often seem to be too basic to be truly valuable to an upper manager. They are seen as being more appropriate for line personnel. Therefore, these tools are "delegatable."

2. The statistical tools such as the Histogram, Scatter Diagram, and Control Chart and sometimes Designed Experiments are often viewed as too technical and therefore appropriate for specialists to use. Therefore, these tools are also seen as "delegatable."

3. The old tools are useful primarily for numerical data. Managers are often faced with verbal data that reflects the "softer" side of business. They frequently need to organize issues, ideas, and words rather than data.

If these views are held, then there is nothing left that is appropriate for these managers to use. Each manager is thus left with no new way of viewing problems. However, the 7 MP Tools seem to be sophisticated enough to warrant their attention, yet simple enough to be mastered in a fairly short time. Most importantly, the 7 MP Tools actually make the job of management understandable and manageable.

The key to learning these tools lies in the following story. A young man carrying a violin case stopped a cab driver in New York City and asked him, "How do I get to Carnegie Hall?" The cab driver answered, "Kid, practice, practice, practice!"

**The Frankel Corporation:
An Application Case Study**

The Company:
Frankel Corporation

Frankel Corporation is an organization in transition. It was founded in 1947 by Martin Franklin and John Kelly, both fresh out of research for a large defense contractor. Marty was a research chemist working in the development of synthetics and John was a mechanical engineer. Both in their early thirties, they saw the opportunities available as subcontractors to their old employer, who was making the change from turning out tanks to producing sedans and coupes for the civilian market. Frankel Corporation was then born in the proverbial garage (Marty's) producing speedometer assemblies. The company quickly grew to 35 employees.

By 1965, Frankel had grown to 80 employees, but its sales had somewhat stagnated. It was then that the founders decided to expand into the exploding plastics and vinyl market. It was a wise decision that enabled Frankel to grow steadily into the 375 employee, $50 million company that it is today.

The Challenge:
Rekindling Innovation and Continuous Improvement
for Survival

Last year the founders, Marty Franklin and John Kelly, decided to retire. They sold all of their interest in the company to the management and hourly workers. The "new" management team found new challenges. Presently, Frankel's business was split 65% automotive and 35% in several other markets. During the last two years, the automotive business had shown signs of dropping off because the products were maturing and few new applications had been developed. Reversing this slide was the number one priority of top management.

Floyd Custer, the new president of Frankel, had been with the company for over 20 years. He had sincere loyalty to Frankel and its people. He believed that the decisions that now needed to be made were the most important ones in the history of the company. He jokingly referred to this as "Custer's Last Stand." In a way, Frankel was indeed surrounded. Foreign competitors had made significant inroads with their major customers. Therefore, Frankel had to not only develop new products but also decide what business it was in.

The Problem:
An Organization That's Frozen in Old Thinking
and Structured to Stay That Way

Custer felt that the following factors made it difficult for the company to make the necessary moves.

- Too much dependence on specialists (e.g., industrial engineers, design engineers, process engineers) and the founders to be innovative.
- Too many departments.
- R & D had become more like process engineering—very little basic research.
- Too little communication with the sales force—felt cut off from the customers.
- Many new ideas in the past have been only half developed and abandoned.
- Structured communications with hourly workers was very sporadic.

Due to these characteristics, Custer saw Frankel as "frozen" in old ways of thinking and behavior. He wanted to "unfreeze" the organization so that the right people could look at the company's products and potential products with a fresh outlook. Furthermore, Custer wanted the most promising ideas to be carried out to their logical conclusion, not simply short-circuited and dropped.

The Solution:
Multi-Level, Cross-Functional Teams
Using the 7 MP Tools to Explore and Develop Possible Paths

Before the founders retired, Custer had attended a seminar on something called the "Seven Management and Planning Tools." He was told to go and was skeptical to say the least. But once he got some practice in the sessions, he found ideas emerging from the process that would never have occurred to him otherwise. He was impressed, but didn't get a chance to try the tools until now.

Custer's first step was to assemble his entire management team for a weekend training/planning session. The group consisted of:

- President (Floyd Custer)
- VP of Operations (Greg Oste)
- VP of Finance and Administration (Liz Gregorian)
- VP of Sales and Marketing (Harry Jackson)
- R & D Engineering Manager (Ravi Khandur)
- QA Manager (Nina Chase)
- Personnel Manager (Fred Cabral)
- 8 First Line Supervisors

Day 1: Review and Practice the 7 MP Tools

The first day, Custer lead the group through the 7 MP Tools, practicing each one with a simple example from the company. At the end of that first day, Liz Gregorian, VP of Finance, commented, "All of the examples used were from manufacturing and I really don't see how these would help in my end of the business." Custer apologized, assured her that they certainly would apply, and promised to focus on administration as well as manufacturing the next day.

Day 2: Applying the 7 MP Tools to Frankel

After a few words of introduction, Custer answered any questions that participants had from the first day. In general, the comments were positive, but there was some frustration over the lack of clear direction at times. Custer agreed that it was sometimes like "being in a place you've been before, not remembering things clearly, realizing you need a map but then discovering that one doesn't exist." He assured the group that it would get easier as they dealt with issues that were more meaningful and meatier to them.

They then split into three groups to use two of the 7 MP Tools, the Affinity Diagram and the Tree Diagram, to deal with two questions:

Morning	What business(es) could/should Frankel be in?
Afternoon	What parts of the administration system are preventing the company from being as effective as it could be?

By the end of the day, these issues had become much more focused. Smaller teams were formed based on specific expertise and interest and were given the assignment to apply the remainder of the 7 MP Tools where appropriate.

The remainder of the case study will follow through on the work of only two of the eight teams that were formed that day. Team A, coordinated by Harry Jackson, VP of Sales and Marketing, has a new product marketing orientation. Specifically, it will focus on designing a marketing program to introduce the public and car manufacturers to a revolutionary children's car safety system called IMP® (Inertial Mass Protection), which was developed by Frankel over the last 14 months. This was chosen as one of the team projects out of more than 60 ideas generated by the group because:

a) This was a totally new product area for the company and therefore inherently risky.
b) It was a high-tech product so there had to be a thorough understanding of what the technology could and could not do. Technical experts needed to be involved.
c) It had to be done right the first time in order to gain market share. If it failed, others who were already in the market would quickly develop a similar product and wipe out Frankel.

Team B, headed by Liz Gregorian, VP of Finance and Administration, is working on the problem of Missed Promised Delivery Dates. This was chosen because it had been a thorn in everyone's side for at least five years. It also showed up as a major issue in a customer survey, which was done for the first time two months ago.

Each of the 7 MP Tools will have a chapter devoted to it with the sequence of each step clearly marked. Each step will be boxed and numbered. In addition, the Frankel Corporation case study materials will be shown step-by-step with **A** indicating the Product Marketing/Child's Safety System Team and **B** indicating the Administrative/ Missed Promised Delivery Dates Team. The remainder of each chapter will include examples of other uses of that tool as well as helpful hints for constructing and interpreting each one.

— *Notes* —

Chapter 1

Affinity Diagram

> ### Definition
>
> *This tool gathers large amounts of language data (ideas, opinions, issues, etc.), organizes it into groupings based on the natural relationship between each item, and defines groups of items. It is largely a creative rather than a logical process.*

The biggest obstacle to planning for improvement is past success or failure. It is assumed that what worked or failed in the past will continue to do so in the future. We therefore perpetuate patterns of thinking that may or may not be appropriate. Continuous Improvement requires that new logical patterns be explored at all times.

An Affinity Diagram (hereafter called an Affinity) is an excellent way to get a group of people to react from the creative "gut level" rather than from the intellectual, logical level. It also efficiently organizes these creative new thought patterns for further elaboration. These thought patterns have seen teams produce and organize more than 100 ideas or issues in 30-45 minutes. Think of how long that task would take using a traditional discussion process. It is not only efficient, however. It encourages *true* participation because every person's ideas find their way into the process. This differs from many discussions in which ideas are lost in the shuffle and are therefore never considered.

Maybe the best way to understand the usefulness of an Affinity is to examine its roots. It was developed in the 1960s as an analytical tool by Jiro Kawakita, a Japanese anthropologist. As an anthropologist, Kawakita studied countless facets of societies and social institutions. He would make detailed notes of all his observations for later analysis. When that time came he was faced with mountains of information that either showed no clear patterns or fell into only the old, familiar theories. Kawakita developed the KJ Method® so that he could accomplish two important ends:

1) Sift through the large volumes of information efficiently.

2) Let truly new patterns of information rise to the surface for closer examination.

This second purpose is related to what is called "right brain thinking." There is a theory that suggests that the brain has two distinct sections or sides. Simply put, it is theorized that the left side deals primarily with the analytical skills while the right side focuses on the creative process. An Affinity harnesses those creative forces that are often present, even if only unconsciously.

When to Use An Affinity Diagram

We have yet to find an issue for which an Affinity has not proven helpful. However, there are applications that are more natural than others. The "cleanest" use of an Affinity is in situations in which:

a) Facts or thoughts are in chaos. When issues seem too large or complex to grasp, try an Affinity to "map the geography" of the issue.

b) Breakthrough in traditional concepts is needed. When the only solutions are old solutions, try an Affinity to expand the team's thinking.

c) Support for a solution is essential for successful implementation.

An Affinity is **not** suggested for use when a problem: 1) is simple, 2) requires a very quick solution.

Typical Uses of an Affinity Diagram

- An automotive supplier of electronics sits down with representatives of its major customer (one of the "Big Three" auto companies) and identifies their key customer demands in four hours using an Affinity Diagram. The president of the supplier company claims that "we learned more about what our customer wanted in four hours than we have in the last four years."

- A Q.A. department in a computer company uses an Affinity Diagram as a basis for a two day off-site meeting to understand all the operating functions expected of it. They restructured their department as a result.

- A major paper manufacturer uses an Affinity Diagram to plan its entire Total Quality Management program. It organizes implementation teams around key issues/header cards identified by the process.

- An automotive company plans its preventive maintenance program using an Affinity Diagram in a one-day process.

Construction of an Affinity Diagram

Assemble the Right Team

The most effective group to assemble to do an Affinity is one that has the necessary knowledge to uncover the various dimensions of the issue. It also seems to work most smoothly when the team is accustomed to working together. This enables team members to speak in a type of shorthand because of their common experiences. Remember. Don't always get the "old gang" together. Include those with valuable input who may not have been involved in the past. Also, keep the team fluid; bring in resource people as needed. There should be a maximum of five to six members on the team.

A The Child Safety IMP System Marketing Team was composed of:
- Harry Jackson, Sales and Marketing Manager
- Ravi Khandur, R & D and Engineering Manager
- Lauren Baker, Production Manager
- Phil Stanowitz, Q.A. Lab Supervisor
- Kathy Santos, Plastics Supervisor
- Michael Dubois, Regional Sales Rep.

B The Missed Promised Delivery Dates Team was composed of:
- Liz Gregorian, VP of Finance/Administration
- Floyd Custer, President
- Tyrone Gomes, Production Supervisor
- Dorothy Matrix, Sales Department Administrative Assistant
- Bob Alteri, Hourly Worker, Shipping Dept.
- Kirk Rosen, Regional Sales Rep.

Phrase the Issue Under Discussion

It seems to work best when vaguely stated. For example, "What are the issues involved in getting top management support for a TQM process?" There should be no more explanation than that, since more details may prejudice the responses in the "old direction." Also, using the word "issues" is helpful because it's neutral. Responses to it could be positive or negative, good or bad, subjective or quantitative, etc. It's this feeling of "anything goes" that you want to capture. Once everyone agrees on the question, place it at the top of a flip chart page/blackboard so that it's visible to the entire team.

A **IMP System Marketing Team**

B **Missed Promised Delivery Dates Team**

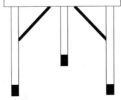

What are the issues involved in successfully marketing the IMP System?

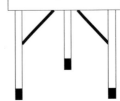

What are the issues involved in missing promised delivery dates?

2 **Brainstorm and Record at Least 20 Ideas or Issues**

Ideas should be generated using the traditional guidelines of brainstorming.

- No criticism of ideas.
- Emphasis on generating a large number of ideas in a short time.
- Participation of every member of the team is encouraged.
- Ideas should be recorded exactly as spoken, not interpreted by the recorder.

Responses can be recorded in one of two ways:

a) Recorded on a flip chart pad while simultaneously transcribed onto small cards (e.g., 1"x3" to 3"x3"), ONE IDEA PER CARD.

b) Recorded directly onto small cards by a recorder or by the contributor without the use of a flip chart.

NOTE 1: <u>Advantages of Flip Chart</u>: Whenever possible we recommend using the flip chart or blackboard. Having the ideas visible to everyone does two things:

a) It stimulates new thoughts as people review the list of ideas periodically.

b) It ensures that ideas are being recorded accurately and not being interpreted by the recorder.

2 **Brainstorm and Record Ideas** (Cont'd.)

NOTE 2: <u>Statement Length</u>: People should be encouraged to be as concise as possible (i.e., no more than five to seven words). On the other hand, one or two word responses should be avoided because they are so open to interpretation (and misinterpretation).

NOTE 3: <u>Statement Structure</u>: Whenever possible the statement should have a noun and a verb. This tends to make the statement less ambiguous.

NOTE 4: <u>Printing</u>: Print legibly and as large as the cards will allow; the cards need to be read from four to five feet away by four to eight people.

NOTE 5: <u>Type of Cards</u>: 3M's Post-it™ Note papers work well when using a smooth printed wall or flip chart as a work surface.

A What Are the Issues Involved in Successfully Marketing the IMP System?

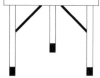

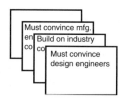

- Must convince design engineers
- Build on industry contacts
- Must convince mfg. engineers in companies
- Must gauge public response
- Need to simulate crash tests
- Don't usually market; fill orders
- First try at non-body components
- No name brand recognition
- Need to know about aftermarket
- First product with electronics
- First major multi-part assembly
- Don't know how much customers willing to pay
- Need potential market size
- Need birth rates
- Don't know how different models need to be
- Need federal guidelines
- Need to market secretly
- Are established producers in the market?
- Need exact cost of necessary auto modification
- Need failure mode analysis
- When to publicize product

- Availability of prototype in adequate numbers
- Need video record of crash tests
- What is cost to auto manufacturer?
- Will auto manufacturer supply market?
- Are there crash situations it won't cover?
- Must meet or exceed federal crash standards
- Full production availability date
- Need predicted life cycle data
- Available for export?
- If exported, licensed or made here?
- What marketing mediums to use?
- How to effectively demonstrate product
- What types of warranties offered?
- Defining its advantages over present product in market

B What Are the Issues Involved in Missing Promised Delivery Dates?

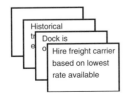

- Hire freight carrier based on lowest rate available
- Dock is overcrowded
- Historical trends of errors
- What is a "shipping error"?
- Time lag in order changes on computer
- When big customer push, switch labels and ship
- Shipping errors only in certain product lines
- Data entry complexity
- Difficult to measure true cost of errors
- High turnover among shippers
- New 11-digit code too long
- Time lag in order changes on computer
- Bar codes damaged/unreadable
- Dock used to store material to be returned to vendors
- How many are paperwork errors—right product shipped?
- Shipping sometimes contacted directly by sales representative to rush orders
- Old shipping boxes easily damaged requiring replacement
- Customer orders still initially handwritten
- How much does each error cost?
- Wrong count by operators on production floor
- Need classification by type of error
- Frequently change freight carriers
- Too slow getting replacement product or paperwork
- Don't tell customer if shipping error known but not detected by customer
- Allow changes to orders over the phone
- Sometimes substitute facsimile product when right product is unavailable
- Computer system too slow — use handwritten forms instead
- Labels fall off boxes
- High turnover among data entry clerks
- No place to segregate customer returns
- Some new, reusable packaging has wrong bar codes
- Lack of training for data entry clerks
- Newest employees go to shipping
- Production bonus system encourages too much speed, not enough accuracy
- Dock is coldest place in winter, hottest place in summer
- How many customers lost that we don't know about?

3 **Sort Ideas Simultaneously into 5-10 Related Groupings**

The teams should mix the cards (like a deck) and spread them out randomly.

NOTE 1: <u>Using Paper/Stock Cards</u>: If using paper or cardboard cards, use a 3 foot by 6 foot table to display the cards. If the finished chart is to be shared with others immediately or transported, put sheets of flip chart paper on the table first. When it's finished simply tape each grouping to the paper.

NOTE 2: <u>Using Self-Adhesive Pads</u>: If using 3M Post-it™ Note paper, work on flip chart paper or a bare table. They don't stick well to a tablecloth or most wallpapers. It's embarrassing when your cards fall off the wall and flutter to the floor like so many autumn leaves.

NOTE 3: <u>Space Needed</u>: Be sure to allow enough space in front of the work surface to enable five or six people to easily see and move the cards.

A and **B** *CHAOS!*

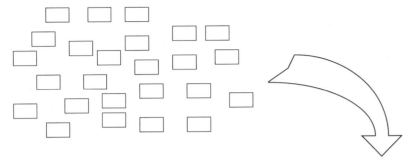

A and **B** *A Team of the Right People…*

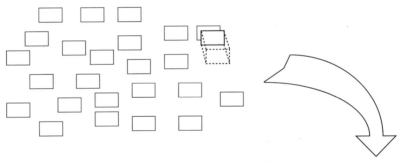

A and **B** *…Finding the Affinity…*

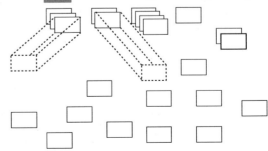

3 **Sort Ideas Simultaneously into 5-10 Related Groupings** (Cont'd.)

At this point the cards should be arranged either by the team or by an assigned individual. This sorting should be done by the entire team simultaneously and in silence.

NOTE 1: <u>Finding First Cut "Affinity"</u>: Although it is possible for one person to complete an Affinity, you lose all of the benefits that come from the melding of perspectives, opinions, and insights. Therefore, I strongly recommend the team approach. Look for two cards that seem to be related in some way. Place those to one side. Look for other cards that either are related to each other or to the original two cards that were set aside. Repeat this process until you have all the cards placed in 6-10 groupings. Do not force-fit single cards into groupings in which they don't belong. These single cards ("loners") may form their own grouping or never find a "home."

NOTE 2: <u>Silent Process</u>: It seems to be most effective to have everyone move the cards at will, without talking. This has two positive results. First it is a sufficiently different experience that "breaks the mold" from the very first step. This seems to encourage unconventional thinking (which is good) while it also discourages semantic battles that can rage on and on (which is bad).

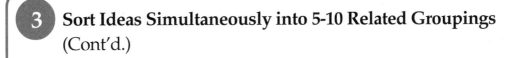

3 **Sort Ideas Simultaneously into 5-10 Related Groupings** (Cont'd.)

NOTE 3: <u>Gut-Level Reaction</u>: Encourage team members not to "contemplate" but to "react" to what they see. Many managers would like to mentally structure all the cards like an oversized chess game. The only thing left to do is to move the pieces to their appointed spots. In an Affinity, speed rather than deliberation is the order of the day. Doing an Affinity should be a high-energy process, not a contemplative exercise.

NOTE 4: <u>Handling Disagreements</u>: Disagreements over the placement of cards should be handled simply and undiplomatically: If you don't like where a card is, move it! It will all eventually settle into consensus (sometimes through exhaustion). This not only speeds up the proceedings, but says it's okay to disagree with your boss by simply moving a card. No muss, no fuss…what a feeling!

NOTE 5: <u>Groupings versus Categories</u>: Do not refer to the columns as "categories." Call them "groupings" instead. This may seem purely semantic, but it seems to keep the team's thinking more flexible while sorting the cards.

3 Sort Ideas Simultaneously into 5-10 Related Groupings (Cont'd.)

NOTE 6: <u>Emergent Thinking vs. Pigeonholing</u>: It is critical that the team allow new groupings to emerge from the chaos of the cards. For the process to work best, members should avoid unconsciously sorting cards into "safe" known categories. This pigeonholing will force fit everything into existing logic, preventing breakthrough from occurring.

A and **B** ...*Among Seemingly Random Ideas!*...

☐ "Loners" ☐

For Each Grouping, Create Summary or Header Cards Using Consensus

Look for a card in each grouping that captures the central idea that ties all of the cards together. This is referred to as a "header" card. This card is placed at the top of each grouping. Many times no such card exists. In these cases, (which happen most of the time), a header card must be created. Gather each grouping together with its header card at the top of the column. The end product looks like a solitaire card game.

NOTE 1: <u>Structure of Headers</u>: The header cards should be, above all, concise. They should state in five to ten words the essence of each grouping. Think of it as an "idea still." Ingredients are thrown into the hopper and distilled until the powerful stuff remains. The header cards should therefore "pack a punch" that would be clear to anyone reading it.

NOTE 2: <u>Constructing a Stand-Alone Header</u>: Imagine that all of the detailed cards under each header card were removed; all that remained were your headers. Would someone who was not a team member understand the <u>essence</u> and <u>detail</u> of the issues raised? This is a good test for the clarity of your header cards.

For Each Grouping, Create Summary or Header Cards Using Consensus (Cont'd.)

NOTE 3: <u>The Two Elements of a Powerful Header Card</u>: Any effective header should:

a) Clearly identify the common thread that ties all of the cards together. This is a central concept like "training." This is not enough, however.

b) Reflect the color and texture of the common thread identified. What are the cards saying about the central concept identified in 'a'? For example, "Provide 'hard'and 'soft' training to all employees."

The header can be a breakthrough idea when it reflects the individual content of the cards as well as the "spirit" of the grouping.

NOTE 4: <u>Avoid Jargon/Cliches</u>: Creating headers is an opportunity to create new twists in old topics. If the headers sound too familiar, they may deserve another look. *Fuzzy words reflect fuzzy thinking*, and *comfortable mindsets beget the familiar.*

A and **B** *Summarize the Theme that Ties Each Grouping Together...*

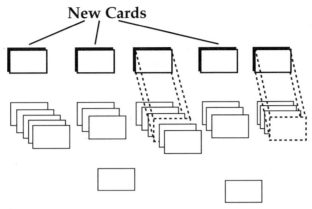

New Cards

---

5 Draw the "Finished" Affinity Diagram

Draw lines around each grouping, thereby clearly connecting all of the items with the header card. Related groupings should be placed near each other and connected by lines. Often when you do this, you find that you must create another header card (referred to as a "super-header") that sums up how these two groupings are related to each other. This would be placed above these two columns and also connected with lines.

This final drawing can be done right on the original sheets or only when the completed diagram has been transferred to another sheet of paper. It is usually transferred because an Affinity Diagram is often shared with people outside the team for comments and changes. Remember that it is a reiterative process that should be changed until it reflects the actual situation and the key factors.

A and **B** ... *And Show the Relationships Among All the Ideas in Final Affinity Diagram*

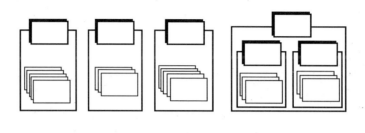

A IMP System Marketing Team Completed Affinity Diagram

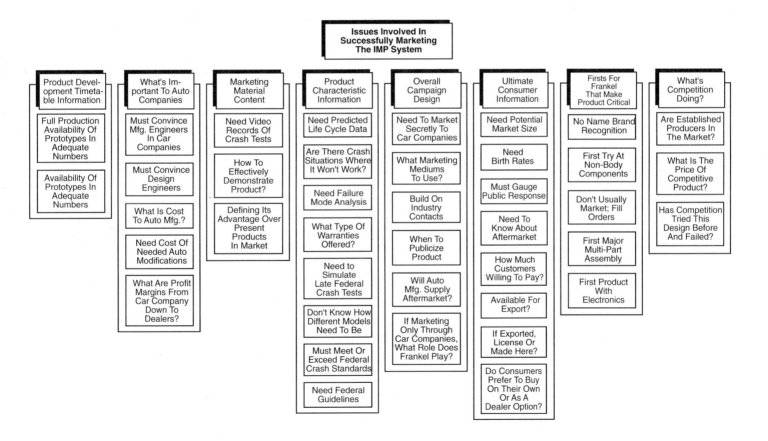

Issues Involved In Successfully Marketing The IMP System

Product Development Timetable Information	What's Important To Auto Companies	Marketing Material Content	Product Characteristic Information	Overall Campaign Design	Ultimate Consumer Information	Firsts For Frankel That Make Product Critical	What's Competition Doing?
Full Production Availability Of Prototypes In Adequate Numbers	Must Convince Mfg. Engineers In Car Companies	Need Video Records Of Crash Tests	Need Predicted Life Cycle Data	Need To Market Secretly To Car Companies	Need Potential Market Size	No Name Brand Recognition	Are Established Producers In The Market?
Availability Of Prototypes In Adequate Numbers	Must Convince Design Engineers	How To Effectively Demonstrate Product?	Are There Crash Situations Where It Won't Work?	What Marketing Mediums To Use?	Need Birth Rates	First Try At Non-Body Components	What Is The Price Of Competitive Product?
	What Is Cost To Auto Mfg.?	Defining Its Advantage Over Present Products In Market	Need Failure Mode Analysis	Build On Industry Contacts	Must Gauge Public Response	Don't Usually Market; Fill Orders	Has Competition Tried This Design Before And Failed?
	Need Cost Of Needed Auto Modifications		What Type Of Warranties Offered?	When To Publicize Product	Need To Know About Aftermarket	First Major Multi-Part Assembly	
	What Are Profit Margins From Car Company Down To Dealers?		Need to Simulate Late Federal Crash Tests	Will Auto Mfg. Supply Aftermarket?	How Much Customers Willing To Pay?	First Product With Electronics	
			Don't Know How Different Models Need To Be	If Marketing Only Through Car Companies, What Role Does Frankel Play?	Available For Export?		
			Must Meet Or Exceed Federal Crash Standards		If Exported, License Or Made Here?		
			Need Federal Guidelines		Do Consumers Prefer To Buy On Their Own Or As A Dealer Option?		

B Missed Promised Delivery Dates Team
Completed Affinity Diagram

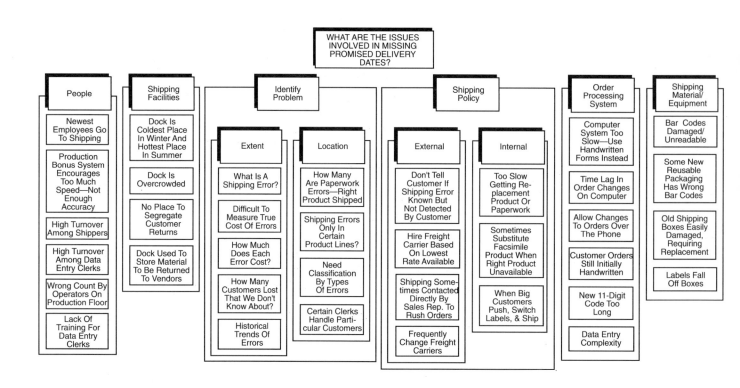

WHAT ARE THE ISSUES INVOLVED IN MISSING PROMISED DELIVERY DATES?

People
- Newest Employees Go To Shipping
- Production Bonus System Encourages Too Much Speed—Not Enough Accuracy
- High Turnover Among Shippers
- High Turnover Among Data Entry Clerks
- Wrong Count By Operators On Production Floor
- Lack Of Training For Data Entry Clerks

Shipping Facilities
- Dock Is Coldest Place In Winter And Hottest Place In Summer
- Dock Is Overcrowded
- No Place To Segregate Customer Returns
- Dock Used To Store Material To Be Returned To Vendors

Identify Problem

Extent
- What Is A Shipping Error?
- Difficult To Measure True Cost Of Errors
- How Much Does Each Error Cost?
- How Many Customers Lost That We Don't Know About?
- Historical Trends Of Errors

Location
- How Many Are Paperwork Errors—Right Product Shipped
- Shipping Errors Only In Certain Product Lines?
- Need Classification By Types Of Errors
- Certain Clerks Handle Particular Customers

Shipping Policy

External
- Don't Tell Customer If Shipping Error Known But Not Detected By Customer
- Hire Freight Carrier Based On Lowest Rate Available
- Shipping Sometimes Contacted Directly By Sales Rep. To Rush Orders
- Frequently Change Freight Carriers

Internal
- Too Slow Getting Replacement Product Or Paperwork
- Sometimes Substitute Facsimile Product When Right Product Unavailable
- When Big Customers Push, Switch Labels, & Ship

Order Processing System
- Computer System Too Slow—Use Handwritten Forms Instead
- Time Lag In Order Changes On Computer
- Allow Changes To Orders Over The Phone
- Customer Orders Still Initially Handwritten
- New 11-Digit Code Too Long
- Data Entry Complexity

Shipping Material/Equipment
- Bar Codes Damaged/Unreadable
- Some New Reusable Packaging Has Wrong Bar Codes
- Old Shipping Boxes Easily Damaged, Requiring Replacement
- Labels Fall Off Boxes

Summary

Throughout this chapter an Affinity is described as the first step in the 7 MP Tools cycle. It's true that it is most powerful when used in conjunction with the remaining techniques, but it has proven to be tremendously useful as a stand-alone tool. It is not an all or nothing proposition. So, if you don't have the time, or the issue is inappropriate for the complete cycle, try an Affinity and see what happens.

Finally, an Affinity may appear to be very structured with very definite rules. This just is not the case. The way your diagram looks may be quite different from the case study and examples provided. The important thing is that participants allow their creative juices to flow and distill them into the key elements which they must address. An Affinity is a "mine of minds" that yields both jewels and junk that can be refined in the remaining tools or through group discussion.

Other Sources of Information

There are other resources that can help you learn about the uses and techniques of constructing the Affinity, or how to *facilitate* the use of the Affinity, as well as other management and planning tools. Several of these resources are:

- *The Memory Jogger*™ *II*
- *Coach's Guide to The Memory Jogger*™ *II*
- *The Coach's Guide* Package
- *The Learner's Reference Guide to The Memory Jogger*™ *II*
- *The Educators' Companion to The Memory Jogger Plus+*®
- *The Memory Jogger Plus+*® Software
- *The Memory Jogger Plus+*® Videotape Series
- *Tools of Total Quality on CD-ROM*

Exhibit A-1
Affinity Diagram

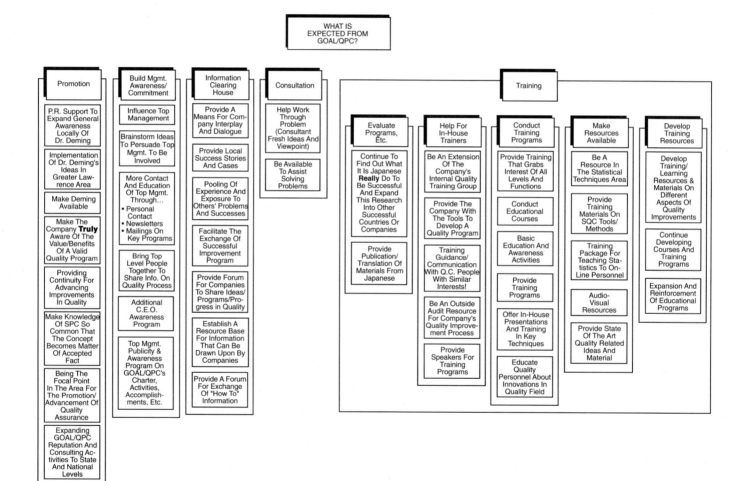

WHAT IS EXPECTED FROM GOAL/QPC?

Promotion

P.R. Support To Expand General Awareness Locally Of Dr. Deming

Implementation Of Dr. Deming's Ideas In Greater Lawrence Area

Make Deming Available

Make The Company **Truly** Aware Of The Value/Benefits Of A Valid Quality Program

Providing Continuity For Advancing Improvements In Quality

Make Knowledge Of SPC So Common That The Concept Becomes Matter Of Accepted Fact

Being The Focal Point In The Area For The Promotion/ Advancement Of Quality Assurance

Expanding GOAL/QPC Reputation And Consulting Activities To State And National Levels

Build Mgmt. Awareness/ Commitment

Influence Top Management

Brainstorm Ideas To Persuade Top Mgmt. To Be Involved

More Contact And Education Of Top Mgmt. Through…
• Personal Contact
• Newsletters
• Mailings On Key Programs

Bring Top Level People Together To Share Info. On Quality Process

Additional C.E.O. Awareness Program

Top Mgmt. Publicity & Awareness Program On GOAL/QPC's Charter, Activities, Accomplishments, Etc.

Information Clearing House

Provide A Means For Company Interplay And Dialogue

Provide Local Success Stories And Cases

Pooling Of Experience And Exposure To Others' Problems And Successes

Facilitate The Exchange Of Successful Improvement Program

Provide Forum For Companies To Share Ideas/ Programs/Progress in Quality

Establish A Resource Base For Information That Can Be Drawn Upon By Companies

Provide A Forum For Exchange Of "How To" Information

Consultation

Help Work Through Problem (Consultant Fresh Ideas And Viewpoint)

Be Available To Assist Solving Problems

Training

Evaluate Programs, Etc.

Continue To Find Out What It Is Japanese **Really** Do To Be Successful And Expand This Research Into Other Successful Countries Or Companies

Provide Publication/ Translation Of Materials From Japanese

Help For In-House Trainers

Be An Extension Of The Company's Internal Quality Training Group

Provide The Company With The Tools To Develop A Quality Program

Training Guidance/ Communication With Q.C. People With Similar Interests!

Be An Outside Audit Resource For Company's Quality Improvement Process

Provide Speakers For Training Programs

Conduct Training Programs

Provide Training That Grabs Interest Of All Levels And Functions

Conduct Educational Courses

Basic Education And Awareness Activities

Provide Training Programs

Offer In-House Presentations And Training In Key Techniques

Educate Quality Personnel About Innovations In Quality Field

Make Resources Available

Be A Resource In The Statistical Techniques Area

Provide Training Materials On SQC Tools/ Methods

Training Package For Teaching Statistics To On-Line Personnel

Audio-Visual Resources

Provide State Of The Art Quality Related Ideas And Material

Develop Training Resources

Develop Training/ Learning Resources & Materials On Different Aspects Of Quality Improvements

Continue Developing Courses And Training Programs

Expansion And Reinforcement Of Educational Programs

Exhibit A-2
Affinity Diagram

MAINTAINING A SUCCESSFUL C.I. PROCESS

Understand Customer Requirements
- Consult Customer
- Interpret Customer Requirements Correctly In Specs & Design
- Provide Operational Definition Of Output
- Identify Customer

Provide Training
- Know Quality Improvement Tools
- Investigate Other Continuous Improvement Efforts

Establish Controls
- Establish Measurement System
- Develop An Effective Corrective Action System
- Make Project By Project Improvement
- Determine Process Capability
- Define Process

Getting Management Commitment
- Involve Middle Mgrs. & Top Mgrs. In Steering Committee
- Establish Consistent Reward Systems
- Provide Job Security: Freedom To Fail
- Provide Middle Managers With Staff Support
- Provide Clear Program Goals
- Provide The Time For Middle Mgrs. To Participate
- Create A Steering Committee With Real Power

Improve Communications In All Areas
- Grant Access To Information
- Employee Involvement
- Break Down Barriers

— *Notes* —

Chapter 2

Interrelationship Digraph

Definition

This tool takes a central idea, issue, or problem, and maps out the logical or sequential links among related items. It is a creative process that shows every idea can be logically linked with more than one other idea at a time. It allows for "multidirectional" rather than "linear" thinking to be used.

In planning and problem solving, it is obviously not enough to just create an explosion of ideas. An Affinity allows some initial organized creative patterns to emerge but an Interrelationship Digraph (hereafter I.D.) lets logical patterns become apparent. By "logical" I mean that you can apply questions like "If I do this, what else among all these items will happen?" or "Does the item result from any of the other ideas generated?" It's therefore more structured than a free-wheeling Affinity that relies on "gut feelings." However, an I.D. is very much like an Affinity in that you must not approach it with a predetermined pattern in mind. This **requires** you to break out of the normal "linear thinking" in which there is a straight line cause and effect "stream" that is neat and tidy. However, most conditions or events are actually caused by a "web" of causes that look more like a Rube Goldberg invention than a predictable path. Therefore, an I.D. requires multidirectional thinking. This is very consistent with the Japanese belief that key ideas and connections rise naturally to the surface like cream in a bottle of milk.

To some, the I.D. process may appear like reading tea leaves. There may be an

Interrelationship

element of truth to this. The Japanese feel very comfortable putting their faith in the "process" to produce the desired results. They think that as long as the process (or ritual) is carried out correctly, then everything else will follow. Most Americans, however, want to see "in the black box" to observe each step as it happens. They want to know the details. The Japanese are generally more comfortable dealing with ambiguous situations. Therefore, sitting back and reserving judgment while the patterns in an I.D. emerge may be "un-American" for many. All I can say is "trust me"—the I.D. <u>process</u> produces unanticipated findings that are worth the wait.

When to Use an Interrelationship Digraph

An I.D. is exceptionally adaptable to both specific operational issues as well as general organizational questions. For example, a classic use of the I.D. at Toyota focused on all the factors involved in the establishment of a "billboard system" as part of their JIT program. On the other hand, it has also been used to deal with issues underlying the problem of getting top management support for TQC.

In summary, an I.D. should be used when:

a) An issue is sufficiently complex that the interrelationship between and among ideas is difficult to determine.

b) The correct sequencing of management actions is critical.

c) There is a feeling that the problem under discussion is only a symptom.

d) There is ample time to complete the required reiterative process involving doing the I.D., reviewing it, modifying it, reviewing it again, etc.

Typical Uses of an Interrelationship Digraph

- A Total Quality Management Steering Committee is not satisfied with the level of participation in project teams. The Steering Committee members feel there are some obvious and more subtle reasons. The president of the company is determined to find the root cause(s) of the problem and eliminate it (or them).

- The president and her direct reports decide to implement a new management compensation system. They realize it is very complex and has an enormous potential impact (good and bad). They need to identify all of the characteristics of a successful system. Just as important, they need to identify the critical few that will kill the system, if not addressed.

- A director of manufacturing is required by a major customer to install a bar code system. He must, within the next quarter, identify and resolve the key bottlenecks in the manufacturing, material handling, warehousing, and shipping operations.

- The store manager of a supermarket has received a 100% increase in customer complaints (from an average of 15/week to 30/week) over the last eight weeks. He needs to prioritize the complaints and look for cause patterns so that the proper course can be taken, e.g., training, new equipment, store set-up.

Construction of an Interrelationship Digraph

 Assemble the Right Team

As is true of an Affinity and the remainder of the tools, the aim is to have **the right people, with the right tools, working on the right problems**. This means that an important step is to define the necessary blend of people in a group of five to six individuals.

NOTE: <u>Team Needs Close Working Knowledge of Issue</u>: The "right" team may or may not be the same people who would construct an Affinity. An I.D. is often more implementation oriented so that it may require people who are closer to a situation than would be needed to do an Affinity.

A and **B** The same team that was assembled for the Affinity is appropriate for the I.D. construction process. This is true because both teams represent a vertical slice of the Frankel Corporation that includes top management, middle management, supervisors, staff support, outside sales, and hourly workers.

2   Agree on the Issue/Problem Statement

Choose the cards that will be part of the I.D.; generate the I.D. cards as needed. **Clearly** define one statement that states the key issue under discussion. This issue and the I.D. cards can be from four sources:

a) The most common source is the Affinity. The Affinity may produce the issue for the I.D. in two ways:
 1. The team can decide by consensus which of the header cards is most critical. In this case, the header card would become the issue for the I.D.

NOTE: Brainstorm Additional Cards: This would require the team to brainstorm additional ideas to "flesh out" the skeleton created in the Affinity. In other words, the team would use the cards in the chosen key Affinity groupings to construct an I.D., combining these with additional ideas that they've generated.

 2. The I.D. can be based upon the same key issue as was the Affinity. In this case, the I.D. would be used to identify the primary issues based on root causes among all the ideas generated and grouped in the Affinity. Used this way, the I.D. provides a means to, as Joseph Juran states, "separate the vital few from the trivial many." Think of the Affinity as a photograph of the United States from space. It reveals a general geography and "lay of the land." Picture the I.D. as a photograph from a high flying aircraft. The territory is

2 — Agree on the Issue/Problem Statement (Cont'd.)

the same, but the details and the relationship among the various landmarks is much clearer.

b) An I.D. is sometimes used as the first step in the process rather than an Affinity. Therefore, an I.D. issue can be any key point for which it's important to find the root cause or bottleneck.

NOTE: When an I.D. Starts the Cycle: If an I.D. is used as the initial tool in the cycle, generate and record the ideas just as in an Affinity.
- Brainstorm
- Use cards
- Post to a flip chart

c) The "effect" statement in a Cause & Effect Diagram can also be the focus of an I.D. In this use, simply take the most basic cause in each "bone" within the 4 M's (Manpower, Methods, Materials, and Machinery) and transfer them to I.D. cards.

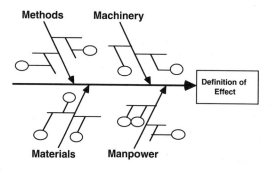

Agree on the Issue/Problem Statement (Cont'd.)

d) A <u>Tree/System Flow Diagram</u> is another source of issues and I.D. cards. This tool, which is covered in depth in Chapter 3, simply takes a key issue or goal and breaks it down into its component parts. It generally takes a broad goal and maps out the means to achieve that goal in greater and greater levels of detail. Use the lowest level of detail as the basis for an I.D.

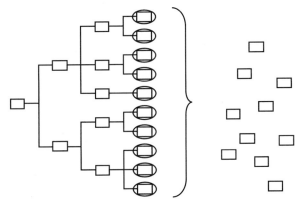

3 Lay Out All of the Ideas/Issue Cards

The team must now distribute the cards on the work surface so that the interrelationships among ideas can be shown.

NOTE 1: <u>Use of Work Surface</u>: The best work surface for an I.D. seems to be flip chart sheets. Tape two sheets together end to end along the length. Either tape these sheets to the wall or place on a table. Another option is to use rolled "butcher paper" which allows you to create sheets of any size without constant taping.

NOTE 2: <u>Use Post-it™ Notes Whenever Possible</u>: Try using the 3" x 3" 3M Post-it™ Notes as cards. They stay in place both on horizontal and vertical work surfaces. If you plan to use them, stay away from open windows and team members with hayfever. Also try using these for the I.D. since it is most efficient to re-use them (once the I.D. has been transcribed onto smaller paper) rather than to create new cards.

NOTE 3: <u>I.D. Without Cards</u>: It's also possible to do an I.D. without cards. You would simply write each idea directly on the flip chart sheets. This may work in some situations, but it tends to become rather messy if someone changes his/her mind about the placement of an idea. An I.D. can look complicated enough without items crossed out, etc. I clearly prefer cards because they're portable, not permanent. If you do choose to

3 Lay Out All of the Ideas/Issue Cards (Cont'd.)

use the "cardless" methods, be sure to place each idea in a box ☐

Place the cards on the work surface using one of three methods:

a) <u>Using the Affinity Groupings as a Model</u>
In this method there are three basic steps:
- Place the Affinity cards, keeping the groupings, in cause and effect sequence as much as possible.
- Place seemingly related groupings next to each other. This will already be done for those cadres under a "superheader." Look for other header cards that may also be related in some way to the "superheader."
- Space the cards to allow for drawing "relationship arrows."

b) <u>Random Distribution</u>: This method is done just as the name suggests—randomly. However, first you must:
- Remove the header cards from the Affinity.

3 **Lay Out All of the Ideas/Issue Cards** (Cont'd.)

- Take the remaining cards and mix them thoroughly. Then lay out the cards randomly on the work surface.

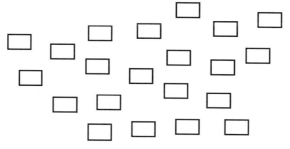

c) <u>One By One</u>: In this method the group places all of the brainstormed cards on the side of the flip chart sheet. (These cards could come from original brainstorming or the Affinity.) Simply choose a card from among the total and place it in the center of the sheet. Then ask the question:

  "Does any other card (among the brainstormed items) either cause or result from this chosen card?"

Once another card is chosen, place it next to the original card and draw an arrow between the two cards going in the desired direction.

3 Lay Out All of the Ideas/Issue Cards (Cont'd.)

Repeat this question again and again until all of the cards are placed in the chart. Each time you place a card in the chart and draw the "relationship arrow," also ask:

"Does this card either cause or result from any of the other cards in the chart?"

Therefore, any time you place a card on the paper, you're not only looking for the <u>primary</u> cause and effect relationship, but also any relationship with any of the cards already on the paper.

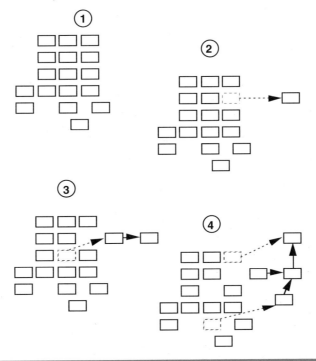

3 Lay Out All of the Ideas/Issue Cards (Cont'd.)

NOTE 1: <u>Ideal Number of Issues/Cards</u>: An I.D. is not appropriate for all instances. Although the Japanese have examples of I.D.'s with over 100 interrelated items, it is most effective when dealing with between 15 and 50 items. If there are fewer than 15 items, the issue is really too simple. If there are more than 50 items, it is generally too complex to be very helpful.

NOTE 2: <u>The Danger of Additional Cards</u>: Unless there is a glaring omission among the cards, try to avoid adding cards to the I.D. If there is one danger to using an I.D., it is a tendency to "explode" into an uncontrollable tangle of issues.

NOTE 3: <u>Spacing Cards</u>: When placing the cards, be sure to leave space between them (at least 1/2") to allow for drawing the relationship arrows.

NOTE 4: <u>Random vs. Pre-Organized Placement of Cards</u>: At this point, it may not seem that it would make a difference if you placed the I.D. cards randomly or based on the Affinity groupings...but it might. Remember that an I.D. is an exercise in "multidirectional thinking" rather than "linear thinking." The "pre-organized" method has the advantage of being neater, more rational. That's part of the problem. It's too rational. It does show interrelationships among groupings, but it encourages linear thinking. On the other hand,

3 Lay Out All of the Ideas/Issue Cards (Cont'd.)

random placement forces you to think in multiple directions because you have no choice. There are no patterns to fall into. There is a saying that applies here, "From confusion comes learning." So if you can put up with the appearance of chaos, the random method usually works best.

NOTE 5:   <u>Amount of Discussion</u>: Unlike an Affinity, for which there is a "no talking" rule, an I.D. requires some discussion about the placement of cards and relationship arrows. Both types of decisions should be made based on the consensus of the team. However, these discussions should be brief so that the process doesn't bog down into fits of hair-splitting. Remember that this is the <u>first</u> version of the I.D. It will be reviewed and revised as needed.

The IMP System Marketing Team opted to construct this I.D. using the Affinity groupings to "pre-organize" the items generated.

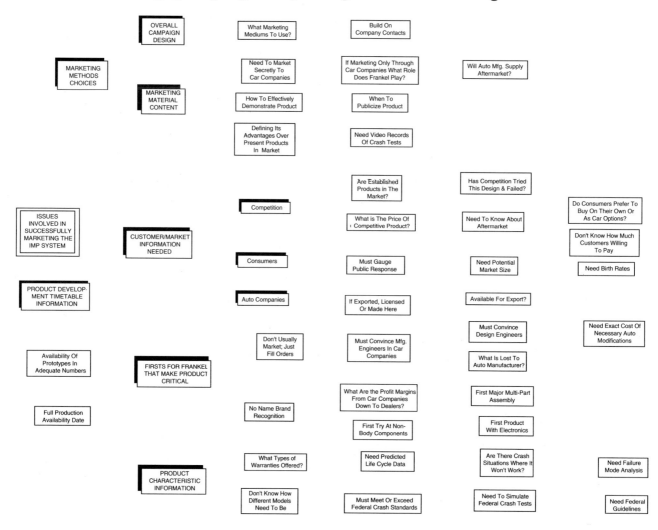

B The Missed Promised Delivery Dates Team chose to use the random method of Affinity card distribution in making their I.D.

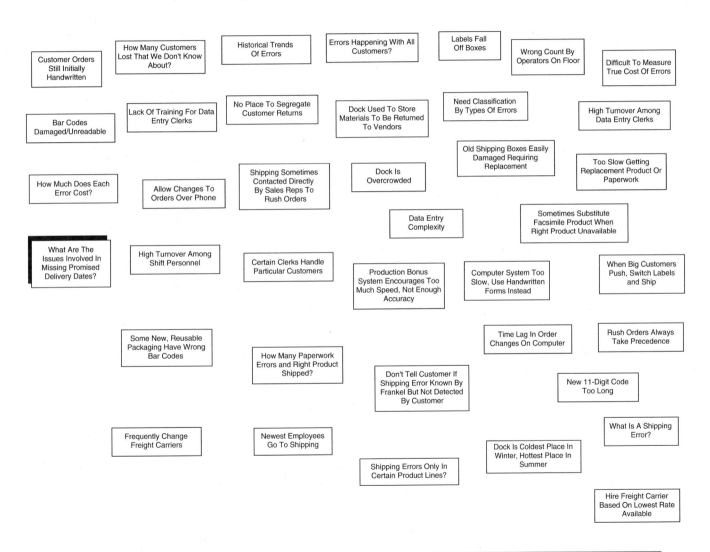

Customer Orders Still Initially Handwritten

How Many Customers Lost That We Don't Know About?

Historical Trends Of Errors

Errors Happening With All Customers?

Labels Fall Off Boxes

Wrong Count By Operators On Floor

Difficult To Measure True Cost Of Errors

Bar Codes Damaged/Unreadable

Lack Of Training For Data Entry Clerks

No Place To Segregate Customer Returns

Dock Used To Store Materials To Be Returned To Vendors

Need Classification By Types Of Errors

High Turnover Among Data Entry Clerks

Old Shipping Boxes Easily Damaged Requiring Replacement

Too Slow Getting Replacement Product Or Paperwork

How Much Does Each Error Cost?

Allow Changes To Orders Over Phone

Shipping Sometimes Contacted Directly By Sales Reps To Rush Orders

Dock Is Overcrowded

Data Entry Complexity

Sometimes Substitute Facsimile Product When Right Product Unavailable

What Are The Issues Involved In Missing Promised Delivery Dates?

High Turnover Among Shift Personnel

Certain Clerks Handle Particular Customers

Production Bonus System Encourages Too Much Speed, Not Enough Accuracy

Computer System Too Slow, Use Handwritten Forms Instead

When Big Customers Push, Switch Labels and Ship

Some New, Reusable Packaging Have Wrong Bar Codes

How Many Paperwork Errors and Right Product Shipped?

Don't Tell Customer If Shipping Error Known By Frankel But Not Detected By Customer

Time Lag In Order Changes On Computer

Rush Orders Always Take Precedence

New 11-Digit Code Too Long

Frequently Change Freight Carriers

Newest Employees Go To Shipping

Shipping Errors Only In Certain Product Lines?

Dock Is Coldest Place In Winter, Hottest Place In Summer

What Is A Shipping Error?

Hire Freight Carrier Based On Lowest Rate Available

Draw the Relationship Arrows

Once all of the cards are laid out, fill in the relationship arrows that indicate what leads to what. When drawing these arrows the "relationship" should be as concrete as possible. When looking at each card you should be asking just one question:

"Which other cards are caused/influenced by this card?"

This question must be repeated for each card until all of the possible interrelationships have been considered. It is critical that this be done for <u>each</u> card in the diagram.

For example: In the Missed Promised Delivery Dates Team's I.D. (see page 59), this question was applied to the card:

"Data Entry Complexity"

The team decided that it caused/influenced five other cards
— Wrong Count by Operators on Floor
— High Turnover Among Data Entry Clerks
— Reusable Containers Have Wrong Bar Codes
— Computer Too Slow, Use Handwritten Forms Instead
— Time Lag in Order Changes on the Computer

NOTE 1: <u>Asking Two-Way Questions</u>: There may be a strong push in a team to ask two questions of each card simultaneously:

Draw the Relationship Arrows (Cont'd.)

"What does this card cause?" and "Which items result in this card?" Experience shows this causes great confusion. If you just repeat the "cause-out" question for each card, then you will identify all the two-way relationships by default.

NOTE 2: <u>Two-Way Arrows</u>: Avoid two-way arrows. Make a decision as to which item is the major influencer. This is often the most valuable outcome of the discussion. In the end, two-way arrows lead to an endless loop that really doesn't provide new information.

NOTE 3: <u>Building Bridges</u>: When one arrow crosses another, build a "bridge," as shown:

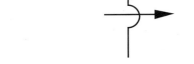

This makes the I.D. easier to read when it's completed.

A The IMP System Marketing Team's I.D. with Relationship Arrows Completed (First Round)

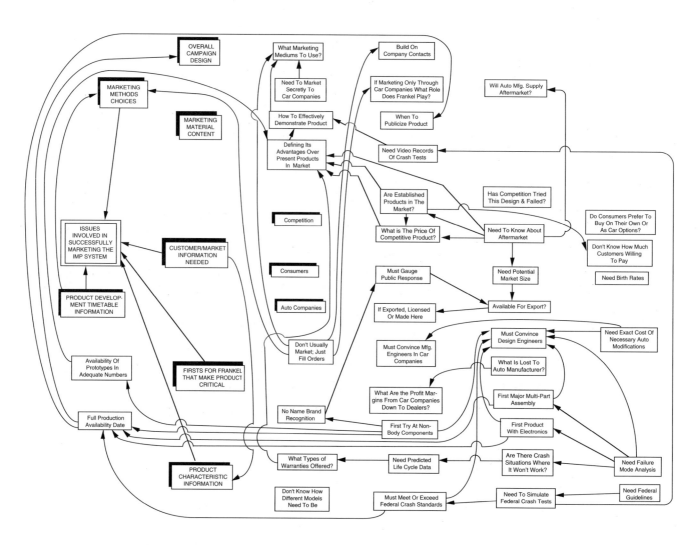

B Missed Promised Delivery Dates Team's I.D. with Relationship Arrows Completed (First Round).

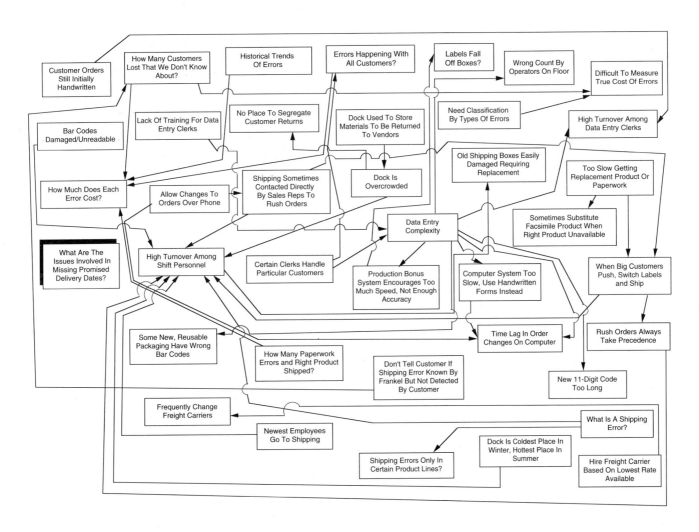

5 Review and Revise the First Round I.D.

At this step the team transfers the I.D. to a paper that can be photocopied and distributed to members of the team and outside resources for review, comments, and changes.

NOTE 1: <u>Suggested Paper Size</u>: An 11" x 17" ledger size paper used lengthwise seems to work best.

NOTE 2: <u>Making Revisions Visible</u>: When drawing arrows during revision, use a different colored pen. This will be of enormous help when you're trying to discuss proposed changes to the I.D. You will notice that the I.D. from the IMP System Marketing Team in this section has certain arrows in bold. These indicate those arrows that were added after it was reviewed by the team.

A The IMP System Marketing Team's I.D. with revisions in bold appears on page 61.

B The Missed Promised Delivery Dates Team's I.D. remained unchanged when reviewed.

A IMP System Marketing Team's I.D. (with revisions in bold)

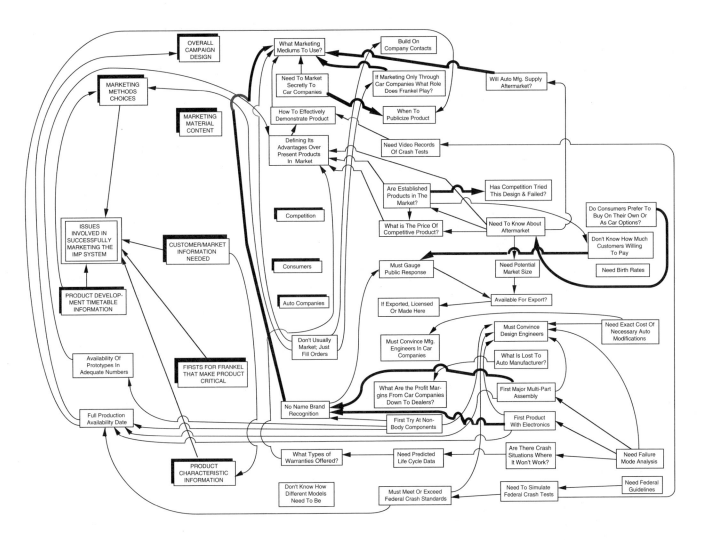

Select Key Items for Further Planning

You've finally arrived at the "tea leaf reading ceremony." At this point you can answer the question, "What does all of this mean?"

You can do this because while you were analyzing whether any two ideas were related, there was a cumulative picture appearing. The "picture" consists of patterns of clustered relationship arrows either converging on or emanating from certain I.D. cards. Therefore, the interpretation process starts simply:

a) First find the I.D. card which has the largest number of arrows either leading to it or coming from it.

b) Continue to find I.D. cards with the next highest number of arrows, the next highest number after that, etc.

c) Review all of the cards in the I.D. (including those already chosen) to find items that have most (or all) of their arrows either coming in or going out.

> <u>Outgoing Arrows Dominant</u>: This indicates a basic cause that, if solved, will have a spillover effect on a large number of items.

> <u>Incoming Arrows Dominant</u>: This may represent a secondary issue or bottleneck that may actually be as important to address as the original item.

6 **Select Key Items for Further Planning** (Cont'd.)

Choose either item depending on how far back in the process you want to go.

d) Look for I.D. cards that may have far fewer relationship arrows clustered around them but that may be key items nonetheless.

e) By consensus confirm that the items with the greatest number of arrows really are the key factors to be tackled. Decide whether there are any other items that don't pass the "arrow test" that should be included. In the interest of manageability, choose a maximum of five to seven key factors to pursue further.

7 Draw the Final I.D., Highlighting the Identified Key Factors

Indicate the key factors by placing the designated cards in a "double hatched box," as shown:

NOTE: <u>Avoiding the Arrows "Numbers Game:"</u> In team's final I.D., notice that there are several "double hatched" boxes that don't pass the "arrow test." In other words, these ideas are shown as key factors whereas other ideas with the same number or more arrows connected to them are not. For example, the item "Time Lag In Order Changes On Computer" has only three arrows connected to it. But the team decided that, based on their own experience, this item deserved further attention. Another example from the same I.D.: It was decided that "Rush Orders Always Take Precedence" was a key idea that only had two arrows connected to it, but was nonetheless at the heart of the problem. Remember, DATA MUST BE USED AS A COMPLEMENT TO EXPERIENCE, NOT AS A RE-PLACEMENT.

A IMP System Marketing Team's Completed
I.D. (key items identified)

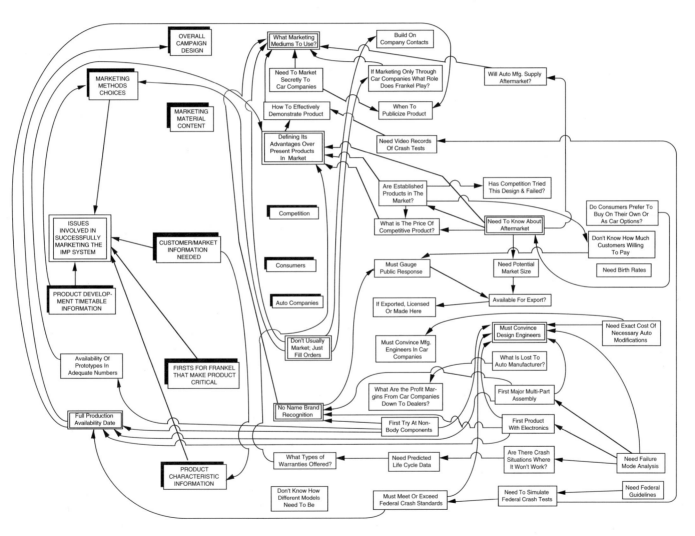

B Missed Promised Delivery Dates Team's Completed I.D. (key items identified)

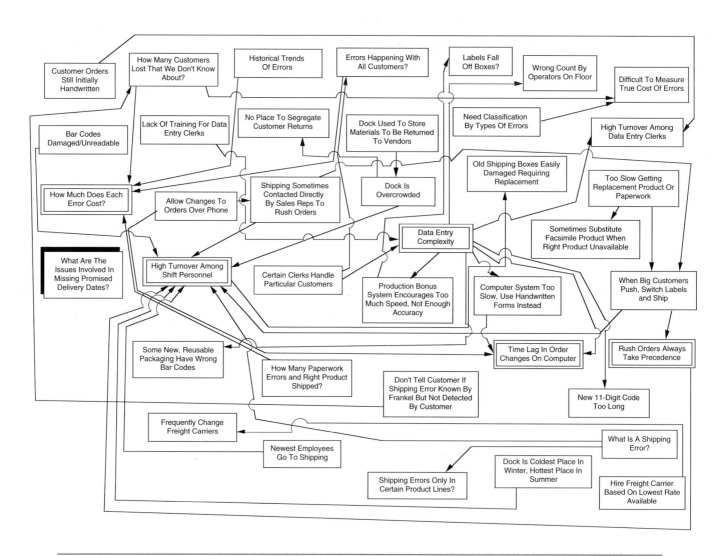

Matrix Model Option

Experience has shown that many people find the I.D. to be graphically chaotic. Many class participants have made very helpful suggestions for streamlining that I.D. concept. Most have been based on some type of matrix format which allows for a methodical item by item comparison. Following is a composite of most of the matrix model suggestions focusing on the issue of "Repeat Service Calls." On page 69 are the same items interrelated using the "traditional" I.D. method. Notice that there are more interrelationships shown in the Matrix Format I.D. This may happen because the matrix <u>forces</u> you to look at each possibility in a more systematic way, whereas the user may get "careless" as the traditional I.D. gets larger and more complex.

The choice of method (traditional vs. matrix) is dependent on the nature of the problem and the teams. In a free-wheeling, creative situation, the traditional I.D. encourages an atmosphere of wide-open, breakthrough thinking that eventually organizes itself. In a situation that calls for a thorough examination of all of the items under discussion, the matrix model seems well suited. Ultimately, the tool must reflect the task and the team. If a team is extremely uncomfortable with the apparent chaos of the traditional I.D., they simply won't use it. In this case, go with the matrix model and preserve the benefits of multidirectional thinking.

Exhibit I-1
I.D. Matrix Model
Issues Involved in Repeat Service Calls

	1	2	3	4	5	6	7	8	9	10	11	12	13	IN	OUT	Total
1. Lack of Good People			↑	←		←		↑		↑				2	3	5
2. Unreasonable Customer					←							←		2		2
3. Wrong Person Sent	←					←	←	←		↑	←		←	6	1	7
4. Lack of Trades Experience in Mgmt.	↑					↑	↑	↑	↑	↑	↑	↑	↑		9	9
5. Lack of Clear Customer Expectations		↑					←	↑	↑			←	↑	2	4	6
6. Lack of Knowledge Matching People To Job Required	↑		↑	←			←	↑	↑		↑		←	3	5	8
7. Lack of Formal Record of What Final Job Is			↑	←	↑	↑		↑	↑	↑	↑		↑	1	8	9
8. Poor Matching of People	←		↑	←	←	←	←		↑	↑	←		←	7	3	10
9. Lack of Clear Job Expectations by Subcontractors			←	←	←	←	←						←	6		6
10. Wrong Tools	←		←	←			←	←						5		5
11. Lack of Information on Job			↑	←		←	←	↑					↑	3	3	6
12. Advertising Promises		↑		←	←									1	2	3
13. Lack of Knowledge of Jobs by Subcontractor Interviewer			↑	←	←	↑	←	↑	↑		←			4	4	8

Exhibit I-2
Interrelationship Digraph
Issues Involved in Repeat Service Calls

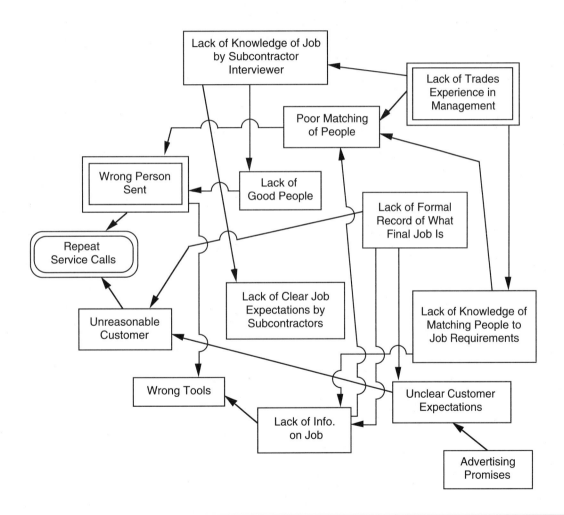

Summary

The Interrelationship Digraph is actually an exercise in "creative logic." These are usually seen as mutually exclusive, but the I.D. finds a way to marry cause and effect thinking with the freedom of brainstorming. In this way it is similar to the Fishbone Chart/Ishikawa Diagram in the 7 Old Tools. However, the Fishbone Chart has no way of clearly showing interrelationships between cause categories. The I.D. provides at least a partial response to the most frequent complaint about the Fishbone: "How do you know what the most important cause is among the many possibilities generated?"

Finally, the I.D. may appear chaotic at times, but it is more reflective of reality than "neater" models. So if you can trust the "ritual," the outcome just may surprise you.

Other Sources of Information

There are other resources that can help you learn about the uses and techniques of constructing the I.D., or how to *facilitate* the use of the I.D., as well as other management and planning tools. Several of these resources are:

- *The Memory Jogger*™ *II*
- *Coach's Guide to The Memory Jogger*™ *II*
- *The Coach's Guide* Package
- *The Learner's Reference Guide to The Memory Jogger*™ *II*
- *The Educators' Companion to The Memory Jogger Plus+*®
- *The Memory Jogger Plus+*® Software
- *The Memory Jogger Plus+*® Videotape Series

Exhibit I-3
Interrelationship Digraph
Issues Involved in Influencing Top Management to Commit to CWQC

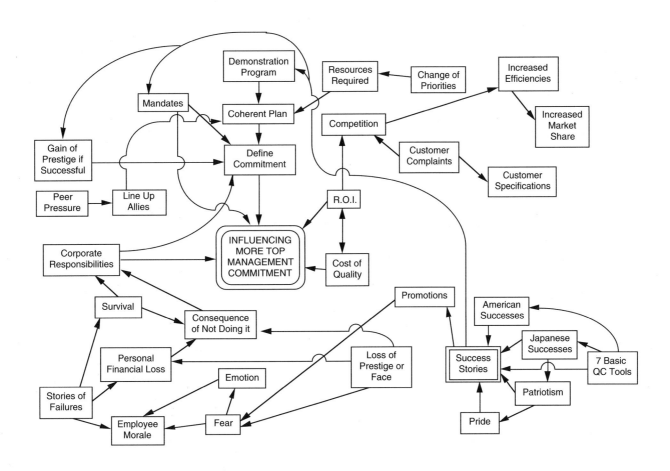

— *Notes* —

Chapter 3

Tree Diagram

Definition

This tool systematically maps out in increasing detail the full range of paths and tasks that need to be accomplished in order to achieve a primary goal and every related subgoal. In the original Japanese context, it describes the "methods" by which every "purpose" is to be achieved. The Tree Diagram brings you from "Motherhood and Apple Pie" objectives to the nitty gritty details of implementation.

With the introduction of the Tree Diagram we enter a more comfortable world: pure linear logic. The appearance of chaos in an Affinity and the Interrelationship Digraph is replaced by an orderly structure that resembles an organization chart placed on its side.

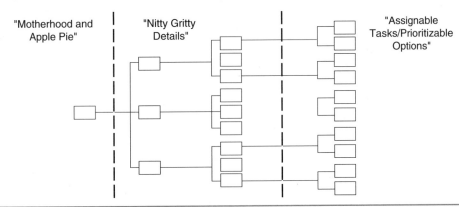

Tree

In many ways an Affinity and Interrelationship Digraph force the key issues to rise to the surface. An Affinity uses gut-level creativity while the Interrelationship Digraph is based upon multidirectional, cause and effect thinking. The Tree Diagram takes these key issues and explodes them down to the lowest practical level of detail.

Key Questions Answered

- What is the sequence of tasks that needs to be completed in order to fully address the key issue/objective/problem?

- What are all of the component parts of the key problem that need to be addressed?

- Does the implementation logic hang together?

- How complex (or simple) will the solution implementation be?

- What are the assignable tasks/prioritizable options that could spin off from the one key issue?

The Tree Diagram creates a systematic focal point for any team that must make sure all bases are covered and that the "lines of logic" are sound. This may appear to be a tedious, mechanical process, but it in fact results in frequent breakthroughs. In both an Affinity and Interrelationship Digraph we sometimes miss entire lines of thought. The Tree Diagram often provides a picture of the issue that makes such gaps obvious.

When to Use the Tree Diagram

- When a specific task has become the focus but is not a simple "assignable job."

- When it is known (or suspected) that implementation will be complex.

- When there are strong consequences for missing key tasks, e.g., safety or legal compliance issues.

- When a task has been considered a simple one yet has run into repeated roadblocks in implementation.

Typical Uses of a Tree Diagram

- A manufacturer of lawn mowers is performing a QFD* study on the customer requirements for a high quality lawn mower. The Tree Diagram enables the manufacturer to break down this broad requirement into specific expressions of what the customer truly wants. This creates a list of customer needs that are all at the same level of detail. This makes it possible in the QFD process for customers to prioritize similar items. It also allows the company to do a competitive analysis on each specific demand. (See page 77 for this example.)

*Quality Function Deployment is a methodology which helps define and identify key customer demands. It then ensures that those demands are met at a competitive cost, using the proper technology, has the needed reliability, and is produced using the necessary quality and production system. It is previously a system of matrices but makes heavy use of the Affinity and Tree Diagrams. The best available reference is *Better Designs in Half the Time* by Bob King, GOAL/QPC, 1988.

- A company is having a problem with declining attendance at its Quality Circle meetings. The Steering Committee constructed a Tree Diagram to define specific preventive and corrective actions that could be taken in order to improve performance. (See page 78 for this example.)

- A company is implementing Total Quality Control (TQC) and must define all of the organizational changes that must be pursued. They define nine major areas that must be addressed in any TQC Plan. (See page 79 for this example. This Tree Diagram is only one of nine produced in this case. "Daily Operational Changes" is the fifth area of change that has been identified and expanded into a Tree Diagram.)

- A Regional Quality and Productivity Center conducts a strategic planning session to build a five year plan. It identifies six broad areas of strategy to pursue. The Marketing example shown is one of those six focus points expanded into a Tree Diagram. (See page 80 for this example.)

- A company faces a lack of support for its continuous quality and productivity process among middle managers. The Steering Committee uses a Tree Diagram to create a list of priority actions that would reward and reinforce middle management support of C.I. (See page 81 for this example.)

Exhibit T-1
Lawn Mower Example

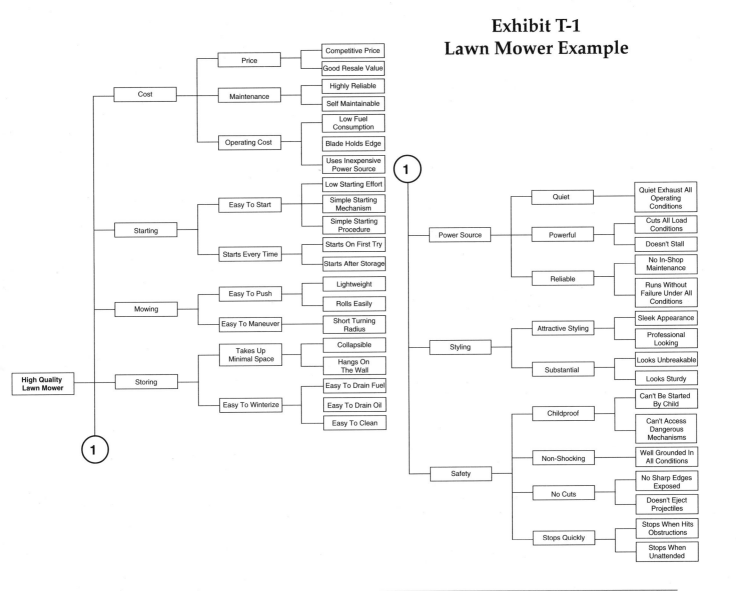

Exhibit T-2
Poor Attendance at Q.C. Meetings

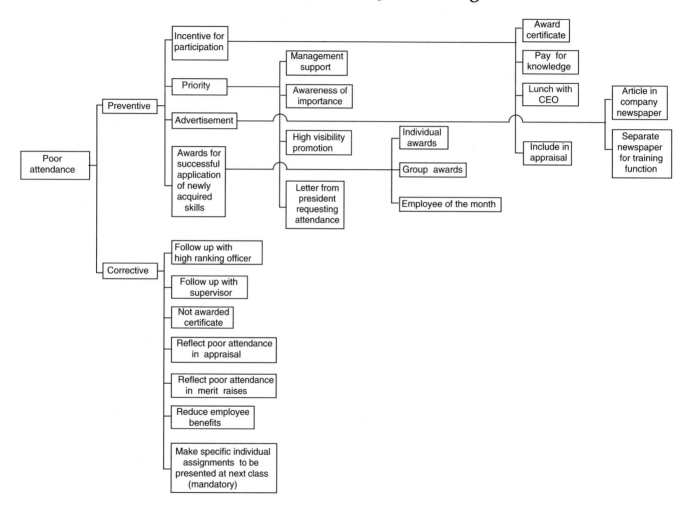

Exhibit T-3
Daily Operational Changes
Needed for a
Continuous Improvement
Program

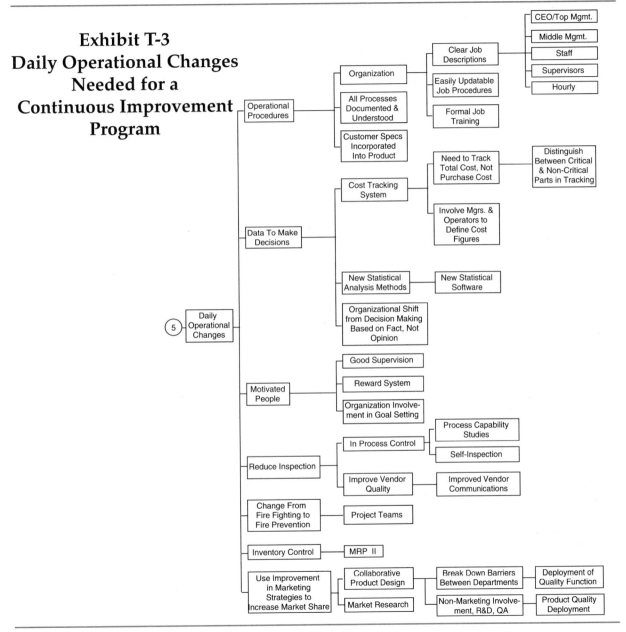

Exhibit T-4
Marketing Tasks for Regional Quality and Productivity Group

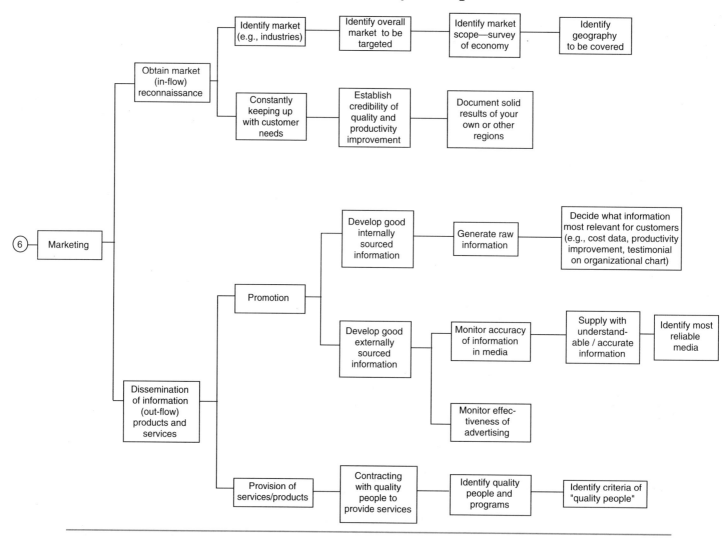

Exhibit T-5
Ways to Deal With the Lack of Management Rewards and Reinforcement for Continuous Improvement

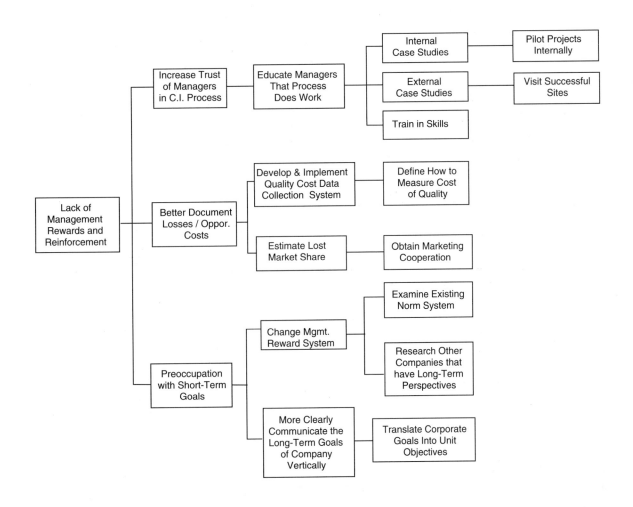

Construction of a Tree Diagram

 1 **Choose the Tree Diagram Goal Statement**

The Tree Diagram starts with an overall goal, which is systematically broken down into the methods by which it will be achieved. The goal statement can come from three sources:

a) <u>The Interrelationship Digraph</u>: If the I.D. is used, then the pattern of arrows will create two types of focus issues for the Tree Diagram:

- Core Cause: This is an I.D. item with most, if not all, arrows coming from it. If this type of issue is resolved, it will have a significant effect on a number of other I.D. issues.

- Secondary Issue: This is an I.D. issue with most arrows going into it. This usually indicates that it is an important item that might actually be a better definition of the original I.D. problem statement.

NOTE 1: <u>When to Choose Either Type of Issue (Core or Secondary)</u>: Which type of issue from the I.D. that you choose for the Tree Diagram (Core/Emerging Arrows or Secondary/Converging Arrows) depends on the insights that either provides. Most of the time, the core issue gets you to the heart of the problem. However, sometimes the secondary issue puts the original problem in a whole new light. Also, the secondary issue has many of the causes already identified. Therefore, if time is of

Choose the Tree Diagram Goal Statement (Cont'd.)

the essence, then much of the new material for the Tree Diagram is already created.

b) <u>The Affinity Diagram</u>: Most commonly, the team will agree on a header card that is most key and then use the cards under it as the basis for a Tree Diagram. However, in certain cases (e.g., absolutely critical strategic business issues), it may make sense to turn the entire Affinity Diagram into a Tree Diagram. This gives you the confidence that the main "branches" of the Tree are truly the key paths to explore.

c) <u>Original Issues</u>: Many times the Tree Diagram can be used as the first step in the planning process. Something has to get done, it's complex, it requires thoroughness and has multiple implementation paths…"tree it out."

Assemble the Right Team

The Tree Diagram marks the point in the process where specialized knowledge is usually required. Instead of thinking in terms of themes or relationships, the Tree requires **action planning**. At this stage you want to generate creative plans, but you want the alternatives to be realistic. Therefore, the teams that generated the Affinity and the I.D. tend to bring in "new blood" who have intimate knowledge of implementation issues.

 The IMP System Marketing Team

The IMP Team generated two different types of issues from its I.D. One had a straight Marketing orientation (Use the Best Mix of Marketing Mediums) while the second issue was rather technical (Convince Design Engineers). Therefore, the original team split into two subgroups and brought in additional resources to continue the problem-solving process:

Marketing Mediums	Design Engineers
Harry Jackson, *Sales & Mktg. Mgr.*	Ravi Khandur, *R&D & Eng. Mgr.*
Michael Dubois, *Reg. Sales Rep.*	Lauren Baker, *Production Mgr.*
+	Phil Stanowitz, *QA Lab Supervisor*
Representatives of: Safety Lab,	Kathy Santos, *Plastics Supervisor*
Existing Advertising Agency	+
	Representatives of: Safety Lab,
	Cost Accounting

B The Missed Promised Delivery Dates Team

The Delivery Dates Team also created two different themes from their I.D. One was a system design item (Reduce Data Entry Complexity) while the other appeared to be a Human Resource/Staffing issue (Reduce Turnover Among Shippers). The team also divided into two subgroups with additional "hands on" people included:

Shipper Turnover	Data Entry
Floyd Custer,* *President*	Liz Gregorian,* *Finance/Admin.*
Dorothy Matrix, *Sales*	Tyrone Gomes, *Production*
+	+
Representatives of:	Representatives of: EDP and
Distribution and HRD	Training & Development

* In a typical process, both the president and the vice president might hand this activity off to functional managers. The problem diagnosis is getting quite focused, so they might not have to be so intimately involved. However, in the case study everyone is trying to master the tools as well so they will stay involved through completion.

Generate the Major Tree Headings

One of the most difficult things to do in constructing a Tree Diagram is to choose the right Tree branches to explore. There is a danger that we will always use the traditional categories. Many complex implementations require new paths of action. In order to prevent the "old line thinking" it has proven to be very helpful to do another Affinity to determine the first level of detail for the Tree Diagram. The header cards would become the first (broadest) level of detail.

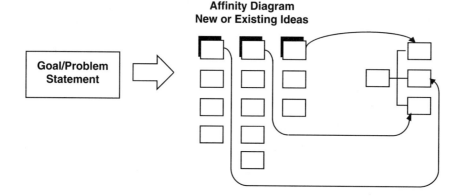

If an Affinity is used, the cards can come from three sources:

- <u>Previous Affinity Diagram</u>: This would usually occur when the chosen problem/goal statement was one of the original header cards. Then the cards under that header would be used as a foundation for the new Affinity.

3 Generate the Major Tree Headings (Cont'd.)

- <u>Previous Interrelationship Digraph</u>: The cards leading into the chosen problem/goal statement would be used as a starting point for generating more brainstormed ideas.

- <u>Open Brainstorming</u>: It's entirely possible that the team could simply start again with an implementation focus.

NOTE 1: <u>Should All the Cards From Previous Affinity or I.D. Be Used Exactly As They're Written</u>: Team members should use previous Affinity or I.D. cards as suggestions of task areas, not always literally. The only reason to use the cards at all is to take advantage of the original creative thinking of the team. The focus has shifted to action/implementation so the cards will often be restated to reflect that shift. For example, an original Affinity card might read "Design Engineers Skeptical." The team might restate it to read "Must Convince Design Engineers of New Mfg. Capacity."

At this point, the team has created a first cut of the tasks/parts of the system that must be addressed in implementation. The major output is the major tree headings (first level detail). The team also has generated the major tree tasks that fall under that first level of detail. However, these tasks are unorganized and usually incomplete. They must now be laid out and systematically analyzed in the "treeing" process.

B Affinity Diagram
Reduce Data Entry Complexity

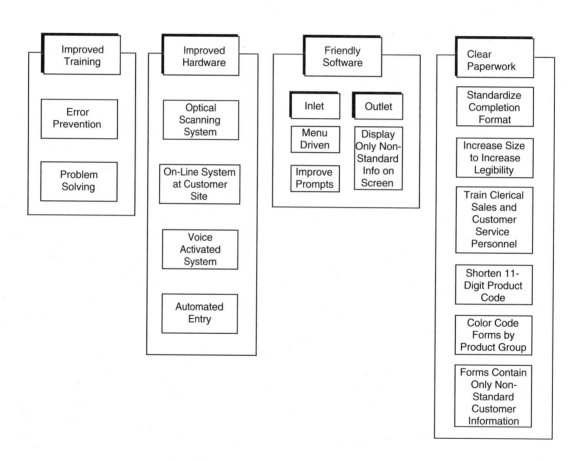

B Affinity Diagram
Reduce Turnover Among Shippers

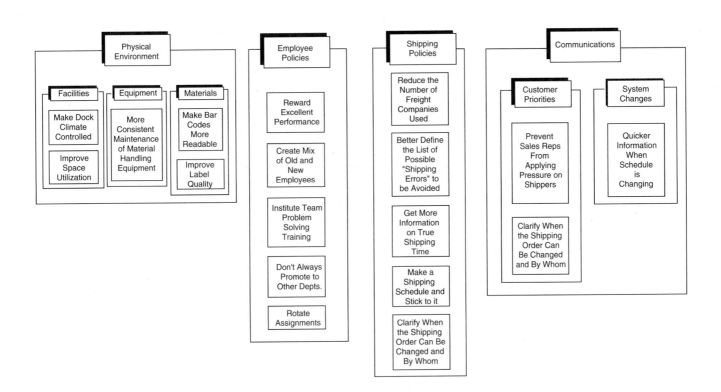

Break Each Major Heading into Greater Detail

a) Place the goal/problem statement to the extreme left of a flip chart, a butcher paper sheet, or a table. If you're using either flip chart or butcher paper, it works best if the paper is laid vertically since the Tree is usually taller than it is wide.

NOTE 1: <u>Cards or Paper Method?</u> Besides the quality of analysis in the Tree Diagram, there is a choice of methods. Many people choose to use a flip chart (or even just a piece of paper for several people) rather than the cards as used in the Affinity and I.D. It appears to be easier, but in the long run it usually isn't, because once items are written, it's difficult to change. The Tree Diagram moves from the most general to the more specific as you move from left to right. During the "Treeing" process, items and tasks often change positions. Cards make it easier to move items as needed. This keeps the process fluid. It must be kept flexible until consensus is achieved. If cards aren't used, one last thought: the team should at least use a flip chart rather than have team members huddled around a small sheet of paper. Visibility seems to encourage creativity.

The remainder of the instructions will assume that cards are used but the same steps would apply if the Tree is drawn directly onto a flip chart, etc.

Break Each Major Heading into Greater Detail (Cont'd.)

b) Ask the question, "<u>What needs to happen/be addressed to resolve/ achieve the problem/goal statement</u>?" Apply this question to each first level Tree item. Review the cards under that item in the Affinity (from Step 3). Choose the task(s) that is most directly related to the first level objective. Remember that as you move from left to right in the Tree the tasks are getting more and more specific. Therefore, the second level of detail should have a direct cause and effect relationship with the first level objectives.

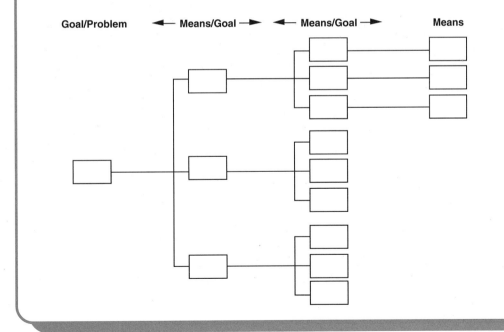

Break Each Major Heading into Greater Detail (Cont'd.)

It must be a direct Tree, because once the Tree Diagram is done, it should reflect all of the tasks to be completed for successful implementation. If there is not a direct cause and effect relationship then something is missing that should be included in the Tree.

c) Repeat the question, "What needs to happen/be adressed . . ." for each level of implementation detail.

If you are using cards from a previous task (Affinity), repeat this question until you have used up all the cards that clearly apply to the Tree.

d) Create new cards to fill in details at all levels in the Tree Diagram.

NOTE 2: <u>Doing a Tree Diagram Without an Affinity</u>: It's entirely possible to do a Tree simply by asking, "What needs to happen/be addressed . . ." repeatedly and creating original cards as you go. Practically speaking, this is the method used most often. As stated earlier, the only risk that you run is that the major paths will be the wrong ones or the "same old stuff." But for more creative applications this may be just what is needed.

A IMP System Marketing Team
Completed Tree Diagram: Convincing
Design Engineers

Must Convince Design Engineers

Prove We Can Provide Completely New Product Group

Provide Performance Data
- Must Meet Or Exceed Federal Crash Standards → Duplicate Federal Crash Tests
- Show It Meets Consumer Durability Standards → Provide Prototypes to Limited Numbers of Very Trusted Employees
- Show It Meets Consumer Cost/Benefit Requirements → Focus Group Studies To Find Price Sensitivity
- Show How Fault Tolerant Product Is In Hands of the Customer → Provide Prototypes to Limited Numbers of Very Trusted Employees

Provide Reliability Data
- Need Failure Mode Analysis → Design Use Array Experiment
- Obtain Reliability Data of Competitive Products → Obtain Competitor Products for Lab Tests
- Show Reliable Use Outside of Laboratory → Provide Competitor Products to Employees For Test / Provide Prototypes to Limited Numbers of Very Trusted Employees

Contact With Our Specialists In Each New Area
- Create A Design Team Approach Among Specialists → Assess the Effect of Each New Product Group on the Others
- Familiarize the Specialists With the Complete Auto Mfrs. Product Line → Hire Specialists in Each New Product Group
- Create a Facility That Will Allow Specialists To "Strut Their Stuff" → Purchase Variety of Cars for Hands-On Design / Create Lab with Sealed Down Production Lines for Joint Design Engineering Meetings

Need Exact Cost of Necessary Auto Modifications

Determine Needed Modifications To Their Products
- Determine All Possible Design Configurations of Auto Company Prods. → Find Out Designs on the Boards for Auto Companies 2-3 Yrs. Out
- Decide On Number of Design Options for Current Product → Design Engineers Produce As Many Working Prototypes As Possible With As Many Designs as Possible / Decide Whether IMP Is Going To Be Proposed Across Entire Auto Prod.Line

Produce Figures in Terms That Are Consistent With Their Cost Acctg. System
- Know Each Auto Company's Cost Accounting System → Hire Cost Accountants On Lg. From Auto Companies / Read Cost Accounting Journals—Look For Auto Company Articles / Ask Auto Companies For Info. On Cost Accounting System

A IMP System Marketing Team
Completed Tree Diagram: Use the Best Mix of Marketing Mediums

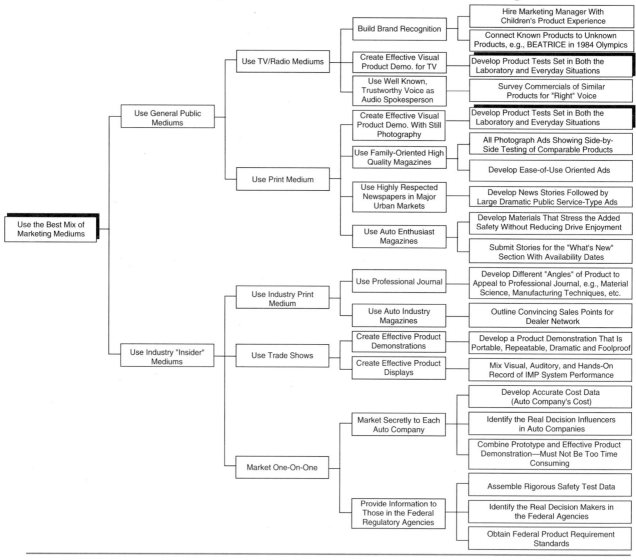

B Missed Promised Delivery Dates Team
Completed Tree Diagram: Reduce Data Entry Complexity

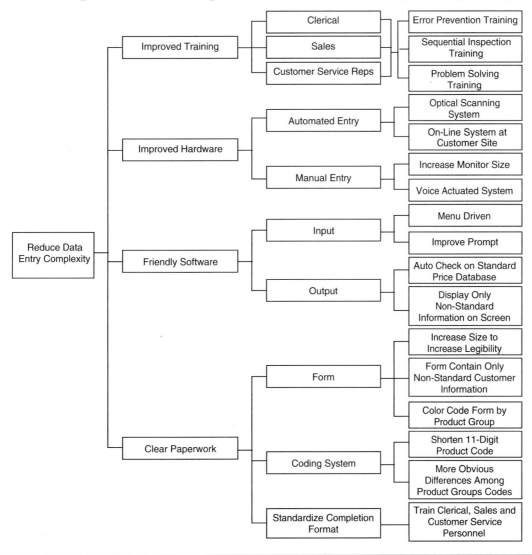

B Missed Promised Delivery Dates Team Completed Tree Diagram: Reduce Turnover Among Shippers

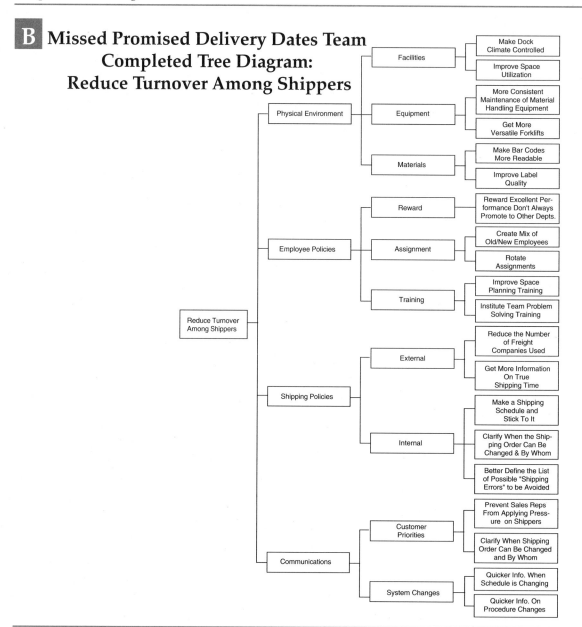

Review the Completed Tree Diagram for Logical Flow and Completeness

The Tree Diagram is effective only if the implementation of each level of detail **really** does accomplish the next higher level of tasks. Therefore, the last step in the process is to apply the question, "Will these actions **actually** lead to these results?" to each level of detail. This checks the specific-to-general (right to left) logic. Likewise, ask the question, "If I want to accomplish these results do I **really** need to do these tasks?" of each level of detail. This confirms the general-to-specific (left to right) logic.

Ideally, this "logic litmus" test should be done by both the team and those closest to the proposed implementation. Be prepared for revisions.

Summary

The Tree Diagram, Affinity Diagram, and Matrix Diagram (Chapter 5) are the most natural and widely used of the Seven Management and Planning Tools. The Tree Diagram, if properly done, reflects the real world of implementing continuous improvement. It forces you to carry rhetoric to its logical conclusions. Ultimately, if it's done well, a Tree Diagram allows you to focus on the smallest details of implementation which make the achievement of the next highest level inevitable.

Remember this, however. The final Tree Diagram maps out the options. It doesn't make evaluations on where you should devote scarce resources. This task remains for the next tool, the Prioritization Matrix.

Other Sources of Information

There are other resources that can help you learn about the uses and techniques of constructing the Tree Diagram, or how to *facilitate* the use of the Tree, as well as other management and planning tools. Several of these resources are:

- *The Memory Jogger™ II*
- *Coach's Guide to The Memory Jogger™ II*
- *The Coach's Guide* Package
- *The Learner's Reference Guide to The Memory Jogger™ II*
- *The Educators' Companion to The Memory Jogger Plus+®*
- *The Memory Jogger Plus+®* Software
- *The Memory Jogger Plus+®* Videotape Series

Chapter 4

Prioritization Matrices

Definition

These tools prioritize tasks, issues, product/service characteristics, etc., based on known weighted criteria using a combination of Tree and Matrix Diagram techniques. Above all, they are tools for decision making.

AUTHOR'S NOTE:

These tools have replaced Matrix Data Analysis as one of the Seven Management and Planning Tools. Matrix Data Analysis is a valuable yet misplaced tool. It should be kept as part of the multivariate analysis process. Its heavy emphasis on rigorous statistical analysis sets it apart from the rest of the tools. Those involved in market/consumer research should use Matrix Data Analysis as an invaluable technique for market segmentation and competitive analysis. Both applications fit well into the steps of the Quality Function Deployment (QFD) process and should be used accordingly. However, if the 7 MP Tools are to be effective, the entire set of techniques must be both useable on a daily basis, yet meaty enough to make the effort worthwhile. The Prioritization Matrices are a helpful addition to an already powerful combination of planning tools.

These are a specialized use of the simplest form (L-Shaped) of the Matrix Diagram, which is covered in depth in Chapter 5. Prioritization Matrices are introduced first because they are often used before broad matrix analysis in a typical cycle using the 7 MP Tools.

Remember, at some point in any planning or problem-solving sequence, progress must be the result of conscious decision making. In other words, someone has to decide what is most important to the organization at that time and then apply those priority considerations to the options under discussion.

Up to this point, the tools have allowed dominant issues and problems to emerge from the process, e.g., the Affinity Diagram and Interrelationship Digraph. They have also helped create the possible courses of action that could address these dominant issues, e.g., the Tree Diagram. The process has looked like this:

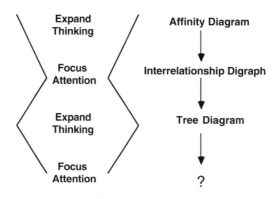

The Affinity Diagram takes specific ideas and allows new general themes to emerge. The I.D. creates focus by identifying bottlenecks and base causes. The Tree Diagram takes any chosen objective (from the Affinity, I.D., or any other source) and breaks it down into its component parts. How, then, do we now focus on choosing the "best" actions and options to pursue Prioritization Matrices?

When to Use the Prioritization Matrices

The traditional "shoot from the hip" planning process has the apparent advantage of creating fewer options but is often the result of "same old way" thinking or pet projects being pushed. Prioritization Matrices are designed to rationally narrow down the focus of any team before detailed implementation planning can happen.

The Prioritization Matrices should be used when:

a) The key issues have been identified and the options generated must be narrowed down.

b) The criteria for a "good" solution are agreed upon but there is disagreement over their relative importance. This applies to the Full Analytical Criteria Method described because it has a technique for ranking/weighting the criteria. The simplified criteria-based matrix on page 121 assumes consensus around criteria ranking without rigorous matrix analysis.

c) There are limited resources for implementation, e.g., time, funds, manpower.

d) The options generated have strong interrelationships.

e) Generating options, not total "laundry lists," all of which have to be done and it is simply a matter of sequencing.

Typical Uses of Prioritization Matrices

- A search committee at a college compares and selects candidates for a key dean's position. The Full Analytical Criteria Method is used to compare the final list of applicants once the first cut has been made based upon a lack of the fundamental requirements (e.g., salary requirements, college education).

- A corporate steering committee for Total Quality uses an Affinity Diagram to generate all of the broad competitive issues that must be addressed within the improvement process. They use the Interrelationship Digraph to identify the most basic organizational weakness. After developing a Tree Diagram they find themselves with an overwhelming number of improvement options. They can't pursue all of them so they submit them for comparison with agreed upon criteria in a series of Prioritization Matrices.

- A planning team for purchasing a major computer system uses the Prioritization Matrices for making the final selection of the best equipment proposals.

- A CEO and his direct reports use Prioritization Matrices to choose appropriate targeted breakthroughs under the Hoshin Planning system.

Construction of Prioritization Matrices

Depending on the complexity of the issue and the time available to prioritize, the construction steps for Prioritization Matrices will vary widely. We will provide instruction for three alternatives:

- The Full Analytical Criteria Method (This is **loosely** based on the work of Thomas L. Saaty. His term for this prioritization process is the Analytical Hierarchy Process as described in his book, *Decision Making for Leaders*, University of Pittsburgh, 1988. This term has been changed to make it consistent with the remainder of the tools and to make it more approachable by mainstream managers.)

- The Consensus Criteria Method

- The Combination I.D./Matrix Method

A **The IMP System Marketing Team** decided to analyze the options for "Using the Best Mix of Marketing Mediums" with the Combination I.D./Matrix Method. The options prioritized were the last level of detail (extreme right hand side) of their Tree Diagram analysis.

B **The Missed Promised Delivery Dates Team** used the Full Analytical Criteria Method on their key issue of "Reducing Data Entry Complexity." The simpler Consensus Criteria Method is used on the same example (Data Entry Complexity) in this chapter for illustrative purposes.

I. The Full Analytical Criteria Method

The Full Analytical Criteria Method is the most complex and rigorous of the Prioritization Matrices. It should be used when:

- The decisions to be made are absolutely critical to the organization.

- There are at least several criteria that must be applied to action options.

- All of the criteria play some significant role in the decision. For example, if one criterion, such as "implementable in the next 30 days," completely overwhelms the other criteria so as to render them unimportant, this complete process is unnecessary and unrealistic.

There are three basic phases in the process:

1. Prioritize and assign weights to the list of criteria.

2. Prioritize the list of options based upon each criterion.

3. Prioritize and select the best option(s) across all criteria.

Steps to Construction

 Agree Upon the Ultimate Goal to Be Achieved

This is simply a confirmation step when the team is using a Tree Diagram to generate the options. The ultimate goal is just the statement at the head of the Tree Diagram as illustrated in Exhibit P-1 that follows on page 106.

Create the List of Criteria to Be Applied to the Options Generated

Assuming that the team working on this issue includes at least some of the ultimate decision makers, this step results from a group discussion. It should be a relatively short process because the team is simply brainstorming the list of criteria, not evaluating the criteria's importance. This will happen in the next step of the process.

NOTE: <u>Wording of Criteria</u>: It's important that each criterion reflects the desired outcome. In other words, the criterion statement should not be neutral. It should be a judgment call. For example, the statement should read "Low cost to implement," not "Implementation cost."

Exhibit P-1
Tree Diagram
Reducing Data Entry
Complexity

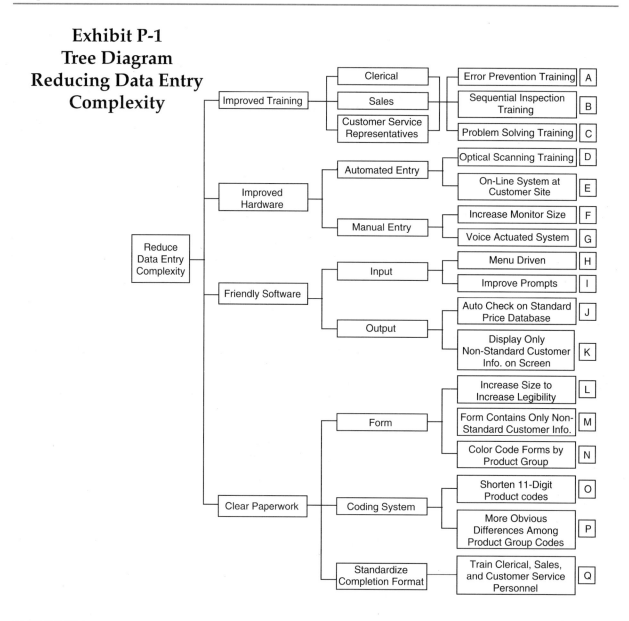

B **The Missed Promised Delivery Dates Team** created the following list of criteria for ranking their Data Entry Complexity options.

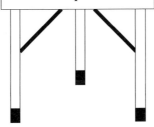

Criteria
- Low Cost to Implement
- No Customized Technology
- Quick to Implement
- Easily Accepted by Users
- Minimal Impact on Other Departments

Judge the Relative Importance of Each Criterion as Compared to Every Other Criterion

Once the total list has been developed, each criterion must be rated so that a weighting number can be assigned.

a) Record the total criteria list on both the vertical and horizontal sides of an L-shaped matrix.

b) Compare the importance of each criterion to every other criterion using the following scale:

1 = Equally Important/Preferred
5 = Significantly More Important/ Preferred
10 = Extremely More Important/ Preferred

1/5 = Significantly Less Important/ Preferred
1/10 = Extremely Less Important/ Preferred

NOTE: <u>Recording the Rating and Its Mirror Image</u>—Whenever you compare any two criteria you should mark that rating in both the row and column of the matrix. For example:

Criteria

		A	B	C	D
	A	1	5		
Criteria	B	1/5	1		
	C			1	
	D				1

Judge the Relative Importance of Each Criterion as Compared to Every Other Criterion (Cont'd.)

When interpreting the matrix, you read across the rows. Therefore, in this example Criterion A is significantly more important than Criterion B. When you read the ratings listed in the next row (B) you now are comparing B to A instead of A to B as in the first row. Because you've already rated that comparison, you must be consistent. The 1/5 rating means that Criterion B is Significantly Less Important/Preferred than Criterion A. Having both ratings recorded allows you to add each row and column and get accurate total rating scores each time.

NOTE: <u>Reading the Ratings</u>: Before adding any rating scores, the scores should be converted to decimal values, e.g., 1/5 = .20 and 1/10 = .10.

c) Add up the rating scores of each column and record the total.

d) Add up the column totals to reach the grand total.

e) Add up each row across the matrix.

f) Divide each row total by the grand total to convert it to a percentage. This percentage becomes the weighting score that will be used as the multiplier in the final matrix that compares all of the options across all of the criteria (Exhibit P-4, page 116).

Exhibit P-2
Full Analytical Criteria Method—Step 1
Ranking the Criteria
Reducing Data Entry Complexity

	Low Cost to Implement	No Customized Technology	Quick to Implement	Easily Accepted by Users	Minimal Impact on Other Depts.	Row Totals (% of Grand Total)
Low Cost to Implement		5	1/10	1/10	1/5	5.4 (.08)
No Customized Technology	1/5		1/5	1/10	1/5	.7 (.01)
Quick to Implement	10	5		1/10	1/5	15.3 (.21)
Easily Accepted by Users	10	10	10		1/5	30.2 (.42)
Minimal Impact on Other Depts.	5	5	5	5		20 (.28)
Column Total	25.2	25	15.3	5.3	.8	71.6 Grand Total

1 Equally Important	
5 Significantly More Important	1/5 Significantly Less Important
10 Exceedingly More Important	1/10 Exceedingly Less Important

Based on the results of the Criteria Ranking Matrix (Exhibit P-2), there is a definite split between two groups of criteria. "Low Cost to Implement" and "No Customized Technology" turn out to be much less important to the final selection process than the other criteria. They are respectively scored at .08 and .01, while the other three criteria have scores of .21, .42, and .28. It is only when there are dramatic differences like this that criteria can be dropped from further analysis. As in the Interrelationship Digraph, the user must not use subjectively generated data to make finely tuned decisions. Just as in the "Arrow Test" of the I.D., the criteria weightings should be used as indicators of major patterns and differences. In this case, "Quick to Implement," "Easily Accepted by Users," and "Minimal Impact on Other Departments" will be used as the criteria against which all of the options under "Reduce Data Entry Complexity" will be compared.

 Compare All of the Options Being Considered to Each Weighted Criterion

Now that the relative importance of each criterion has been established, each option must be judged based on how completely it meets each of the three chosen criteria.

a) Record the total list of options on both the vertical and horizontal sides of an L-Shaped Matrix.

NOTE: Comparing Options to a Criterion: All of the steps described are repeated for each criterion. This example would therefore generate three separate matrices. Only one criterion analysis is shown here for the sake of simplicity.

Compare All of the Options Being Considered to Each Weighted Criterion (Cont'd.)

b) Compare each option to every other option in relation to the criterion being applied. As in Step 3, the scale for comparison is as follows:

1 = Equally quick, acceptable by users, etc.

5 = Much **quicker** implementation, more acceptable by users, etc.

10= Significantly **quicker** implementation, most acceptable by users, etc.

1/5 = Much **slower** implementation, less acceptable by users, etc.

1/10 = Significantly **slower** implementation, least acceptable by users, etc.

These are applied as appropriate to each criterion. Also as in Step 3, each time a comparison score is entered in a row, a corresponding value must be entered in the column. (See example on page 108.)

c) Add the rating scores of each column and record the total.

d) Add up the column totals to reach the grand total.

e) Add up each row across the matrix.

f) Divide each row total by the grand total to convert it to a percentage.

Exhibit P-3
Full Analytical Criteria Method—Step 2
Ranking Options By Individual Criteria
Reducing Data Entry Complexity

Quick To Implement	A	B	C	D	E	F	G	H	I	J	K	L	M	N	O	P	Q	Row Totals (% Grand Total)
A Error Prevention Training	▓	1/5	1/5	5	10	1/5	10	1/5	1/10	1/5	1/5	1/5	1/5	1/10	5	5	1/10	36.9 (.04)
B Sequential Inspection Training	5	▓	5	10	10	1/5	10	1/5	1/5	1/5	1/5	1/5	1/10	1/10	10	5	1/5	56.6 (.07)
C Problem Solving Training	5	1/5	▓	5	10	1/5	10	1/5	1/10	1/5	1/5	1/5	1/5	1/10	5	5	1/10	41.7 (.05)
D Optical Scanning System	1/5	1/10	1/5	▓	1	1/10	5	1/5	1/10	1/5	1/5	1/10	1/10	1/10	1	1/5	1/10	8.9 (.01)
E On-Line System at Customer Site	1/10	1/10	1/10	1	▓	1/5	5	1/10	1/10	1/5	1/10	1/10	1/10	1/10	1	1/5	1/10	8.6 (.01)
F Increase Monitor Size	5	5	5	10	5	▓	10	5	5	5	5	1	1	1/5	5	5	1/5	72.4 (.08)
G Voice Activated System	1/10	1/10	1/10	1/5	1/5	1/10	▓	1/5	1/10	1/5	1/5	1/10	1/10	1/10	1/5	1/10	1/10	2.2 (.00)
H Menu Driven	5	5	5	5	10	1/5	5	▓	1	1	1	1/5	1/5	1/10	5	5	1/5	48.9 (.06)
I Improve Prompts	10	5	10	10	10	1/5	10	1	▓	5	5	1	5	1	10	5	1	89.2 (.10)
J Auto Check On Standard Price Data Base	5	5	5	5	5	1/5	5	1	1/5	▓	1/5	1/5	1/5	1/5	5	5	1/5	42.4 (.05)
K Display Only Non-Standard Info On Screen	5	5	5	5	10	1/5	5	1	1/5	5	▓	1/5	1	1/5	5	5	1/5	53.0 (.06)
L Increase Size To Increase Legibility	5	5	5	10	10	1	10	5	1	5	5	▓	5	1/5	10	5	1	83.2 (.10)
M Forms Contain Only Non-Standard Cust. Info.	5	10	5	10	10	1	10	5	1/5	5	1	1/5	▓	1/5	5	5	1/5	72.8 (.08)
N Color Code Forms By Product Group	10	10	10	10	10	5	10	10	1	5	5	5	5	▓	10	10	1	117.0 (.13)
O Shorten 11-Digit Product Code	1/5	1/10	1/5	1	1	1/5	5	1/5	1/10	1/5	1/5	1/10	1/5	1/10	▓	1	1/10	9.9 (.01)
P More Obvious Difference Among Prod. Grp. Codes	1/5	1/5	1/5	5	5	1/5	10	1/5	1/5	1/5	1/5	1/5	1/10	1	1	▓	1/10	23.2 (.03)
Q Train Clerical Sales & Customer Service Pers.	10	5	10	10	10	5	10	5	1	5	5	1	5	1	10	10	▓	103.0 (.12)
Column Totals:	70.8	56	66	102.2	117.2	14.2	130	34.5	10.6	37.6	28.7	10	23.6	3.9	88.2	71.5	4.9	869.9 Grand Total

1 Equally Quick Implementation
5 Much Quicker Implementation   1/5 Much Slower Implementation
10 Significantly Quicker Implementation   1/10 Significantly Slower Implementation

* These criteria chosen for comparison with each option were taken from the criteria ranking (Exhibit P-2). When there is a significant difference between groups of criteria, the cluster with the highest rating can be used for further analysis. In this case, the top three criteria were chosen.

One could interpret from this chart that the quickest options to implement when compared to all other choices are:

- Color Code Forms by Product Group (.13)
- Train Clerical, Sales, and Customer Service Personnel (.12)
- Improve Prompts (.10)
- Increase Size to Increase Legibility (.10)
- Increase Monitor Size (.08)
- Forms Contain Only Non-Standard Customer Information (.08)

These form a cluster of top priorities only based on the criterion of "Quick to Implement." These results must now be combined with the same comparisons based on the other two criteria.

5   Compare Each Option Based on All Criteria Combined

a) Record all the options on the vertical side of an L-Shaped Matrix.

b) Record all of the criteria that were used as a basis for option comparisons on the horizontal side of the matrix.

c) Transfer the percentage scores from Exhibit P-3 (for all three criteria) under each criterion column.

d) Multiply each percentage score (in each criterion) by the weighted score developed in Exhibit P-2.

5 **Compare Each Option Based on All Criteria Combined** (Cont'd.)

e) Add together the score of each option in every criterion.

f) Add the composite score from each option to create a grand total.

g) Convert each option score into a percentage by dividing the score by the grand total.

Exhibit P-4 shows clear grouping among options. The composite priorities are:

- Color Code Forms by Product Group (.12)
- Improve Prompts (.10)
- Menu Driven (.09)
- More Obvious Differences Among Product Group Codes (.09)
- Increase Size to Increase Legibility (.08)
- Increase Monitor Size (.07)
- Shorten 11-Digit Product Code (.07)

This process has produced a set of seven priorities to be pursued.

Exhibit P-4
Full Analytical Criteria Method Matrix—Step 3
Ranking Options by All Criteria
Reducing Data Entry Complexity

Options / Evaluation Criteria	Quick to Implement	Easily Accepted By Users	Minimal Impact on Other Depts.	Row Totals (% of Grand Total)
A Error Prevention Training	.04 X .21 = .008	.03 X .42 = .013	.03 X .28 = .008	.029 (.03)
B Sequential Inspection Training	.07 X .21 = .015	.04 X .42 = .017	.02 X .28 = .006	.038 (.04)
C Problem Solving Training	.05 X .21 = .011	.04 X .42 = .017	.03 X .28 = .008	.036 (.04)
D Optical Scanning System	.01 X .21 = .002	.03 X .42 = .013	.02 X .28 = .006	.021 (.02)
E On-Line System at Customer Site	.01 X .21 = .002	.01 X .42 = .004	.03 X .28 = .008	.014 (.02)
F Increase Monitor Size	.08 X .21 = .017	.06 X .42 = .025	.08 X .28 = .022	.064 (.07)
G Voice Activated System			.04 X .28 = .011	.011 (.01)
H Menu Driven	.06 X .21 = .013	.09 X .42 = .038	.11 X .28 = .031	.082 (.09)
I Improve Prompts	.10 X .21 = .021	.09 X .42 = .038	.11 X .28 = .031	.090 (.10)
J Auto Check on Standard Info on Screen	.05 X .21 = .011	.06 X .42 = .025	.05 X .28 = .014	.050 (.06)
K Display Only Non-Standard Info On Screen	.06 X .21 = .013	.05 X .42 = .021	.06 X .28 = .017	.051 (.06)
L Increase Size to Increase Legibility	.10 X .21 = .021	.06 X .42 = .025	.09 X .28 = .025	.071 (.08)
M Forms Contain Only Non-Standard Cust. Info.	.08 X .21 = .017	.06 X .42 = .025	.02 X .28 = .006	.048 (.05)
N Color Code Forms by Product Code	.13 X .21 = .027	.13 X .42 = .055	.11 X .28 = .031	.113 (.12)
O Shorten 11-Digit Product Code	.01 X .21 = .002	.12 X .42 = .050	.03 X .28 = .008	.060 (.07)
P More Obvious Differences Among Prod. Group Codes	.03 X .21 = .006	.10 X .42 = .042	.13 X .28 = .036	.084 (.09)
Q Train Clerical Sales and Customer Service Personnel	.12 X .21 = .025	.03 X .42 = .013	.04 X .28 = .011	.049 (.05)
Column Total	.211	.421	.279	Grand Total .911

II. Consensus Criteria Method

There is no denying that the Full Analytical Criteria Method is complete and complex. For critical strategy issues it might be an absolutely essential decision-making tool. However, there are many situations that are both simpler and less critical yet require prioritization and application of criteria.

The Consensus Criteria Method looks very much like the summary matrix (Exhibit P-4) in that it:
- is a simple L-Shaped Matrix
- contains the options as the vertical side
- lists the criteria on the horizontal side

The Consensus Criteria method, however, is different from the Full Analytical method in some very important ways:

- The criteria are weighted simply by the consensus of the team. For example, the least important criterion would be given a value of .10 while the most important one would be given a value of .40 to indicate that the team approximated that it was about four times as important as the lowest rated criterion.

- The options are rank ordered as a group, not based on a systematic comparison of each option to every other option. This rank ordering can be accomplished by:

  - Wide-open consensus
  - Any rating scheme
  - Nominal Group Technique (NGT)

Steps to Construction

1 Construct an L-Shaped Matrix Combining the Options and Criteria to Be Applied

Based upon the lowest level of detail of a Tree Diagram (see Exhibit P-1), list the options to be prioritized on the vertical side of the matrix.

Create the list of appropriate criteria to be applied to the options and place them on the horizontal side of the matrix.

2 Prioritize the Criteria

Through consensus, prioritize the chosen criteria by agreeing on an importance weighting for each item.

NOTE: <u>Recommended Prioritizing Process</u>: When seeking consensus it is often helpful to take a first cut analysis to establish the general areas of agreement/disagreement. The Nominal Group Technique is a popular method for this purpose. The NGT process can be used in the following modified format:

Prioritize the Criteria (Cont'd.)

a) Each person in the group lists the criteria on a sheet of paper

      A.   Quick to Implement
      B.   Easily Accepted by Users
      C.   Minimal Impact on Other Departments
      D.   Low Cost
      E.   Use Available Technology

b) Each person rank orders the five choices by distributing the value 1.0 among the five criteria

      A.   .30
      B.   .30
      C.   .20
      D.   .15
      E.   .05

c) The weightings for each criterion are combined to arrive at a composite ranking score

Criteria	Person #1		Person #2, etc.			Total Across the Entire Team
A. Quick to Implement	.30	+	.30	+...	=	1.85
B. Easily Accepted	.30	+	.30	+...	=	1.25
C. Minimal Impact	.20	+	.10	+...	=	.70
D. Low Cost	.15	+	.20	+...	=	.75
E. Available Technology	.05	+	.10	+...	=	.45

Prioritize the Criteria (Cont'd.)

d) Review each criterion for consistency of weighting among team members. This will help focus further discussion on only those items around which there is widespread disagreement.

Rank Order the Options Based on Each Criterion

With a large number of options to be rank ordered, it is necessary to use some sort of structured ranking process. The most popular method for dealing with this number of items again is the Nominal Group Technique (NGT). The steps to this process would be the same as above in Step 2. However, instead of using weighted ratings, the options would be simply rank ordered under each criterion.

For example, ranking options based upon how **quickly** (or slowly) they can be implemented.

Options	Person #1		Person #2, etc.			Total & Ranking (#) Across the Entire Team
A. Error Prev. Trng.*	6	+	4	+...	=	26 (5)
B. Seq. Insp. Trng.	10	+	8	+...	=	42 (8)
C. Prob. Solvg. Trng.	7	+	6	+...	=	33 (6)
⋮	⋮		⋮			⋮
Q. Train Clerical...	3	+	6	+...	=	30 (4)

* See Tree Diagram Exhibit P-1

Compute the Individual Importance Score for Each Option Under Each Criterion

Calculate the Individual Importance Score of each option by multiplying the option rank number by the criterion weighting number as decided by the team.

For Example:

Options \ Evaluation Criteria	(1.85) Quick To Implement
Error Prevention Training	$5^*(1.85)^+ = 9.25$
Sequential Inspection Training	$8\ (1.85) = 14.80$

* This team option rank number was calculated through the Nominal Group Technique illustrated on page 120.
+ This criterion weighting number was calculated through the Nominal Group Technique illustrated on page 119.

5 ## Compute The Total Ranking Scores Across All Criteria

Once the Individual Importance Score has been calculated for all options under each criterion, add each score together. The option with the highest total score across all the criteria becomes the highest priority.

For Example:

Options / Evaluation Criteria	Quick to Implement (1.85)	Easily Accepted (1.25)	Minimal Impact (.70)	Low Cost (.75)	Available Technology (.45)	Total
Error Prevention Training	5 (1.85) =9.25	3(1.25)=3.75	14(.70)=9.8	4(.75)=3.00	16(.45)=7.20	33.0

Exhibit P-5
Consensus Criteria Method
Reducing Data Entry Complexity

Evaluation + Criteria / Options *	Quick to Implement (1.85)	Easily Accepted (1.25)	Minimal Impact (0.70)	Low Cost (.75)	Available Technology (.45)	Total
(A) Error Prevention Trng.	5(1.85)=9.25	3(1.25)=3.75	14(0.70)=9.8	4(0.75)=3.00	16(0.45)=7.20	33.00
(B) Sequential Inspection Training	8(1.85)=14.8	4(1.25)=5.00	8(0.70)=5.60	6(0.75)=4.50	5(0.45)=2.25	32.15
(C) Problem-Solving Training	6(1.85)=11.1	5(1.25)=6.25	6(0.70)=4.20	9(0.75)=6.75	7(0.45)=3.15	31.45
(D) Optical Scanning Syst.	3(1.85)=5.55	2(1.25)=2.50	4(0.70)=2.80	3(0.75)=2.25	4(0.45)=1.80	14.90
(E) On-Line System at Customer Site	2(1.85)=3.70	1(1.25)=1.25	3(0.70)=2.10	2(0.75)=1.50	1(0.45)=0.45	9.00
(F) Increase Monitor Size	12(1.85)=22.20	11(1.25)=13.75	11(0.70)=7.70	12(0.75)=9.00	11(0.45)=4.95	57.60
(G) Voice Activated System	1(1.85)=1.85	3(1.25)=3.75	2(0.70)=1.40	1(0.75)=0.75	2(0.45)=0.90	8.65
(H) Menu Driven	15(1.85)=27.75	14(1.25)=17.50	13(0.70)=9.10	15(0.75)=11.25	3(0.45)=1.35	66.95
(I) Improve Prompts	17(1.85)=31.45	15(1.25)=18.75	15(0.70)=10.50	16(0.75)=12.00	17(0.45)=7.65	80.35
(J) Auto Check on Standard Price Data	7(1.85)=12.95	9(1.25)=11.25	7(0.70)=4.90	7(0.75)=5.25	6(0.45)=2.70	37.05
(K) Display Only Non-Standard Info	9(1.85)=16.65	7(1.25)=8.75	10(0.70)=7.00	8(0.75)=6.00	8(0.45)=3.60	42.00
(L) Increase Size to Increase Legibility	13(1.85)=24.05	12(1.25)=15.00	12(0.70)=8.40	13(0.75)=9.75	12(0.45)=5.40	62.60
(M) Forms Contains Only Non-Standard Customer Info.	10(1.85)=18.50	10(1.25)=12.50	1(0.70)=0.70	10(0.75)=7.50	9(0.45)=4.05	43.25
(N) Color Code Forms by Product Group	16(1.85)=29.60	17(1.25)=21.25	16(0.70)=11.20	17(0.75)=12.75	15(0.45)=6.75	81.55
(O) Shorten 11-Digit Product Code	11(1.85)=20.35	16(1.25)=20.00	9(0.70)=6.30	11(0.75)=8.25	10(0.45)=4.50	59.40
(P) More Obvious Difference Among Prod. Grp. Codes	14(1.85)=25.90	13(1.25)=16.25	17(0.70)=11.90	14(0.75)=10.50	13(0.45)=5.85	70.40
(Q) Train Clerical, Sales, & Customer Serv. Pers.	4(1.85)=7.40	6(1.25)=7.50	5(0.70)=3.50	5(0.75)=3.75	14(0.45)=6.30	28.45

* 1) The highest score possible is equal to the total number of options
(e.g., 17 = meets criteria most completely, 1 = least meets the criteria). ⟩ From Nominal Group Technique
2) All options are rank ordered.

\+ 1) The highest score possible is equal to the total number of criteria
(e.g., 5 = most important, 1 = least important).
2) In the example "Quick to Implement" = most important and
"Available Technology" = least important.

III. Combination I.D./Matrix Method

Often there is a need to prioritize options based on the elimination of root problems or bottleneck issues, rather than on applying known criteria. This need is well suited to the use of a tool such as the Interrelationship Digraph (I.D.).

The I.D. is a very powerful technique for uncovering unknown patterns of cause and effect. However, it does not distinguish the strength of the cause and effect relationship. An arrow with a "possible" connection is valued as much as one that is "absolutely positively" connected.

The combination I.D./Matrix Method is designed to overcome this problem by showing both the direction of influence and the strength of that influence.

Steps to Construction

 Construct an L-Shaped Matrix Comparing All of the Options to Themselves
Based upon the lowest level of detail of a Tree Diagram (see Exhibit P-6, page 125), list the options to be prioritized on both the vertical and horizontal sides of the matrix.

NOTE: Using Coding Rather Than Full Descriptions: Rather than listing the description of all the options on both the vertical and the horizontal, you can simply number the option descriptions on the vertical and include only the reference numbers (minus descriptions) on the horizontal.

Exhibit P-6
Tree Diagram: Use the Best Mix of Marketing Mediums

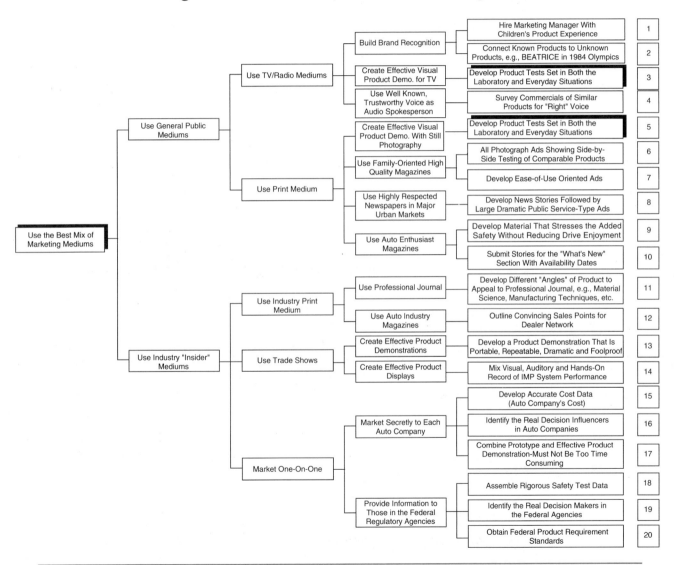

		Hire Marketing Manager With Children's Product Experience	1
	Build Brand Recognition	Connect Known Products to Unknown Products, e.g., BEATRICE in 1984 Olympics	2
Use TV/Radio Mediums	Create Effective Visual Product Demo. for TV	Develop Product Tests Set in Both the Laboratory and Everyday Situations	3
	Use Well Known, Trustworthy Voice as Audio Spokesperson	Survey Commercials of Similar Products for "Right" Voice	4

Use General Public Mediums

Create Effective Visual Product Demo. With Still Photography	Develop Product Tests Set in Both the Laboratory and Everyday Situations	5
Use Family-Oriented High Quality Magazines	All Photograph Ads Showing Side-by-Side Testing of Comparable Products	6
	Develop Ease-of-Use Oriented Ads	7
Use Highly Respected Newspapers in Major Urban Markets	Develop News Stories Followed by Large Dramatic Public Service-Type Ads	8
Use Auto Enthusiast Magazines	Develop Material That Stresses the Added Safety Without Reducing Drive Enjoyment	9
	Submit Stories for the "What's New" Section With Availability Dates	10

Use Print Medium

Use the Best Mix of Marketing Mediums

Use Industry "Insider" Mediums

Use Industry Print Medium

Use Professional Journal	Develop Different "Angles" of Product to Appeal to Professional Journal, e.g., Material Science, Manufacturing Techniques, etc.	11
Use Auto Industry Magazines	Outline Convincing Sales Points for Dealer Network	12

Use Trade Shows

Create Effective Product Demonstrations	Develop a Product Demonstration That Is Portable, Repeatable, Dramatic and Foolproof	13
Create Effective Product Displays	Mix Visual, Auditory and Hands-On Record of IMP System Performance	14

Market One-On-One

Market Secretly to Each Auto Company	Develop Accurate Cost Data (Auto Company's Cost)	15
	Identify the Real Decision Influencers in Auto Companies	16
	Combine Prototype and Effective Product Demonstration-Must Not Be Too Time Consuming	17
Provide Information to Those in the Federal Regulatory Agencies	Assemble Rigorous Safety Test Data	18
	Identify the Real Decision Makers in the Federal Agencies	19
	Obtain Federal Product Requirement Standards	20

1   Construct an L-Shaped Matrix Comparing All of the Options to Themselves (Cont'd.)

	1	2	→	20
1. Hire Marketing Mgr. with Children's Product Experience				
2. Connect Known Product to Unknown Products e.g., Beatrice in '84 Olympics				
↓				
20. Obtain Federal Product Requirement Standards				

2   Compare Each Option to Every Other Option to Determine Direction and Strength of Influence

Review and mark each option on the vertical side of the matrix by asking two questions:

"Does ___(each option)___ cause/influence any of the other options to occur?"

"Where there is a causal relationship, what is its strength?"

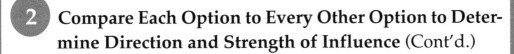

2 **Compare Each Option to Every Other Option to Determine Direction and Strength of Influence** (Cont'd.)

NOTE 1: <u>Asking One Cause Question At A Time</u>: When reviewing each option to find if it is causing or making possible any other option, **don't** simultaneously ask what that option is being caused by. This gets too confusing. Ultimately, all of the options will have arrows coming in and going out as needed. But this results from asking one question at a time in a disciplined manner.

NOTE 2: <u>Mirror Image Arrows</u>: As in the Matrix Model I.D. the total number of In and Out arrows will be added up across each horizontal row. At the same time, the split between In and Out arrows will be calculated. In order to do this, a matching arrow must be placed in the column every time a cause arrow is drawn in the row. They must be pointing in opposite directions, e.g., if the row arrow is pointing up, the column arrow must be pointing to the side.

NOTE 3: <u>Relationship Symbols</u>:

◎ = Strong Relationship
○ = Medium Relationship
△ = Possible/Weak Relationship

2 **Compare Each Option To Every Other Option to Determine Direction and Strength of Influence** (Cont'd.)

NOTE 4: <u>Matching Relationship Symbols</u>: Like the influence arrows, whenever a symbol is entered in a row it must also be entered in the corresponding column. Unlike the arrows, the same symbol is entered.

	1	2	3	4	5
1) Hire Marketing Manager with Children's Product Experience	■		↑◎	↑◎	↑○
2) Connect Known Products to Unknown Products , e.g., Beatrice in '84 Olympics		■			
3) Develop Product Tests Set in Both the Lab and Everyday Situations	←◎		■		
4) Survey Commercials of Similar Products for "Right" Voice	←◎			■	
5) All Photograph Ads Show Side-by-Side Testing Comparable Prods.	←○				■

Tabulate the Relationship Arrows and Relationship Strengths Across Each Row

a) Add the total number of "In" Arrows (those arrows pointing to the left) and enter under the "Total In" column.

b) Add the total number of "Out" Arrows (those arrows pointing upward) and enter under the "Total Out" column.

c) Combine the "Total In" and "Total Out" numbers in the "Total In and Out" column.

d) Add up the values for all of the relationship symbols across the row and enter under the "Strength" column.

 Interpret Chart and Choose the Highest Priority Option

In interpreting the Combination I.D./Matrix Method, you must maintain the balance between direction of influence and strength of relationship. Therefore, rather than looking at any one number to make final selections, try the following sequence:

a) First look at the **Strength** column. This should be the first indicator since a high priority option should have significant, not trivial,

Interpret Chart and Choose the Highest Priority Option (Cont'd.)

relationships with other options, regardless of the number of arrows leading to or from that option. Just use it as a starting point to separate out clusters of options, e.g., highest, medium, low. Work with the highest cluster in examining the second layer of investigation: the Total In and Out.

b) Use the **Total In and Out** to again identify clustering among options that were highly rated, based on strength of relationship. To have an option identified in both categories would surface options that had the most powerful impact on the greatest number of options. Look first for those options that have a <u>significantly</u> greater number of total arrows.

c) Next, look for a dominant "In" or "Out" direction among the options that had high Strength and Total Arrow Scores.

- **Predominantly Out** indicates a base cause. These options would have an impact on a large number of other options. This is often the most expedient short-term activity to pursue.

- **Predominantly In** indicates an outcome or a secondary issue that is often very close to the original goal that was initially raised. It is often a measure of long-term success of the overall plan. It pulls all the pieces together because many options lead to its attainment.

Interpret Chart and Choose the Highest Priority Option (Cont'd.)

- **Split evenly In and Out** indicates approximately equal converging and emerging arrows. Do not get too precise when interpreting patterns. It is not an exact science.

NOTE: <u>A Second Look at Even Splits</u>: When forced with an even split (or close to it) between incoming and outgoing arrows, you could examine the strength of the relationship among the arrows going in each direction. For example, an option may have four arrows going out and five arrows going in, but closer examination shows that all four arrows going out are very strong for a total of 36. On the other hand, the five incoming arrows have either some relationship or a possible one, for a total of only 13.

A **The IMP System Marketing Team** chose eight options to pursue out of a total of 19 original possibilities. They are:

- Develop Product Test Set in Both the Laboratory and Everyday Situations
- Obtain Federal Product Requirements Standards
- Hire Marketing Manager with Children's Product Experience
- Mix Visual, Auditory, and Hands-On Record of IMP System Performance
- All Photographs Show Side-by-Side Testing of Comparable Products

- Develop Ease-of-Use Oriented Ads
- Develop Production Demonstration That Is Portable, Repeatable, Dramatic, and Foolproof
- Outline Convincing Sales Points for Dealer Networker

This list is a combination of analysis and experience. In most of the cases, they passed both the arrows and strength test. Several choices reflected the experience of the group alone. None of the decision rules/indicators should be used exclusively.

Exhibit P-7
Combination I.D./Matrix Method
Use the Best Mix of Marketing Mediums

Legend:
- ◎ (9) = Strong Influence
- ○ (3) = Some Influence
- △ (1) = Weak/Possible Influence

Row Item	Total In	Total Out	Total In and Out	Strength
Hire Mktg. Manager with Children's Prod. Experience	0	6	6	32
Connect Known Products to Unknown Products. e.g. BEATRICE in 1984 Olympics	6	2	8	20
Develop Product Tests Set in Both the Laboratory and Everyday Situations	2	9	11	67
Survey Commercials of Similar Prods. for "Right" Voice	5	3	8	30
All Photograph Ads Show Side-by-Side Testing Comparable Prod.	2	7	9	25
Develop Ease-of-Use Oriented Ads	6	5	11	55
Develop News Stories Followed by Large Dramatic Public Service Type Ads	4	3	7	31
Develop Material That Stresses the Added Safety Without Reducing Drive Enjoyment	4	2	6	42
Submit Stories for the "What's New" Section with Availability Dates	5	1	6	26
Dev. Differ. Angles of Prod. to Appeal to Prof. Journals, e.g., Material, Science, Mfg., Techniq.	0	2	2	10
Outline Convincing Sales Points for Dealer Network	11	3	14	84
Dev. Prod. Demo That is Portable, Repeatable, Dramatic & Foolproof	5	4	9	53
Mix Visual, Auditory and Hands-On Record of IMP System Performance	5	5	10	54
Develop Accurate Cost Data (Auto Company's Cost)	0	1	1	9
Identify the Real Decision Influencers in Auto Companies	1	0	1	1
Combine Prototype & Effective Product Demo. - Must Not Be Too Time Consuming	5	0	5	33
Assemble Rigorous Safety Test Data	3	3	6	34
Identify the Real Decision Makers in the Federal Agencies	0	2	2	10
Obtain Federal Product Requirements Standards	1	7	8	46

Summary

In the end, prioritization is going to happen in any implementation, either by design or default. Those responsible for implementation may play a hunch, take a vote, mandate, analyze for impact, etc., but somehow they will decide what should be done first. This chapter has suggested two approaches:

- Where criteria are known and can be compared (Full Analytical and Consensus Criteria Methods)

- Where relationships and cause and effect among options are most important (Combination I.D./Matrix)

Whichever method you choose to use, the key is to avoid arbitrary priorities that have little or nothing to do with achieving breakthroughs. At this stage of the process you've come too far to fall back into "same old way" thinking. Always leave yourself open to "emergent thinking" in which the unexpected can rise to the surface for consideration. These methods are just one more piece in the creativity puzzle.

Other Sources of Information

There are other resources that can help you learn about the uses and techniques of constructing the Prioritization Matrices, or how to *facilitate* the use of the Prioritization Matrices, as well as other management and planning tools. Several of these resources are:

- *The Memory Jogger™ II*
- *Coach's Guide to The Memory Jogger™ II*
- *The Coach's Guide* Package
- *The Learner's Reference Guide to The Memory Jogger™ II*
- *The Educators' Companion to The Memory Jogger Plus+®*
- *The Memory Jogger Plus+®* Software
- *The Memory Jogger Plus+®* Videotape Series

Chapter 5

MATRIX DIAGRAM

> ## Definition
>
> *This tool organizes large numbers of pieces of information such as characteristics, functions, and tasks into sets of items to be compared. By graphically showing the logical connecting point between any two or more items, a Matrix Diagram can surface which items in each set are related. Beyond the existence or absence of a relationship, it can also code each relationship to show its strength and the direction of the influence.*

Of the tools discussed thus far (Affinity Diagram, Interrelationship Digraph, Tree Diagram), the Matrix Diagram has enjoyed the widest use and is the most versatile. It is based on the principle that whenever a series of items are placed in a line (horizontal) and another series of items are placed in a row (vertical) there will be intersecting points that might indicate a relationship. Furthermore, the Matrix Diagram features highly visible symbols that indicate the strength of the relationship between the items that intersect at that point. Thus, the Matrix Diagram is very similar to the other tools in that new cumulative patterns of relationships emerge, based on the interaction between individual items. In effect, a matrix forces you to consider the correlation between only two items at a time. The two questions are also the same:

1) Is there any relationship between these two items?
2) If yes, what is the strength of that relationship?

Even in this most logical process, unforeseen patterns "just happen."

Matrix

When to Use the Matrix Diagram

- When "motherhood and apple pie" has evolved into definable and assignable tasks that must be "deployed" to the rest of your organization.

- When the "focused activities" generated must be tested against other things that your organization is already doing.

- When your organization is trying to prioritize present activities given new priorities, i.e., choose the present system(s) that helps achieve the greatest number of new objectives.

- When there is a need to get a cumulative numerical "score" that allows you to compare any one item to any other item or all of the other items combined.

Typical Uses of a Matrix Diagram

- **Allocating Organization Tasks:** A training and consulting firm must determine the best allocation of all its organization's "jobs" across each staff member. The complete list of the tasks is compiled and compared to all staff members one by one. Each job/staff member interaction is examined and it's decided whether that person is involved in that task and if so, what is his/her level of responsibility. This is completed individually and then agreed upon by the entire staff. (See page 141 for this example.)

- **Quality Function Deployment/A-1/House of Quality:** The lawn mower manufacturer from the Tree Diagram chapter (see page 77) compares the customer demands that were organized by the Tree Diagram in a QFD matrix one by one to known competitive analysis criteria (e.g., rate of importance, present level of performance, level up). Once this process identifies key customer demands, "Substitute or Counterpart Quality Characteristics" (the internal system elements that, if controlled, will help meet the key customer demands) are created for each demand and placed in the A-1/House of Quality Matrix. The strength of the relationship (if any) between each demand and all of the Substitute Quality Characteristics is determined and recorded in the A-1/House of Quality Matrix. This identifies the most important elements of the internal system that must be controlled or improved if the key customer demands are to be satisfied. (See page 143 for the example.)

- **Evaluating Product/Service Design Options:** A design team uses a matrix to choose which of the child safety system concepts that they've generated is most promising. They first create a Tree Diagram of customer demands and then brainstorm all of the design possibilities that can meet those demands. The matrix is then used to compare each option to each demand. This surfaces the most promising concept by identifying the demands that each concept relates to. This relationship can be either positive (helps meet demand) or negative (works against the demand). (See page 145 for this example.)

- **Developing a Comprehensive TQC Training Process:** A company uses a matrix to look at four critical elements of any training program. First, what are all the possible topic areas that should somehow be covered in support of Total Quality Control (TQC)? Secondly, who should be trained in each subject area? Third, at what depth should each identified trainee be trained? Fourth, who should provide the training in each topic? The matrix allows you to graphically display all of this information on one sheet. (See page 149 for this example.)

Matrix Diagram Formats

The primary reason for the widespread use of the Matrix Diagram is its flexibility. First of all, the content that is chosen for any matrix is limitless. Secondly, there are at least five standard matrix formats that allow you to show relationships among two, three, or four sets of variables in either two dimensions (showing relationships between only two individual variables at a time) or three dimensions (showing the relationship among three individual variables simultaneously). Following are the most commonly used matrix formats.

Most Common Matrix Diagram Formats

1. L-Shaped Matrix

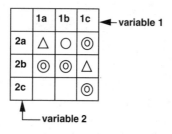

2. T-Shaped Matrix

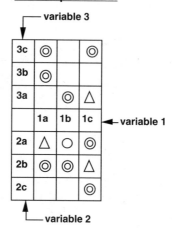

Most Common Matrix Diagram Formats (Cont'd.)

3. Y-Shaped Matrix

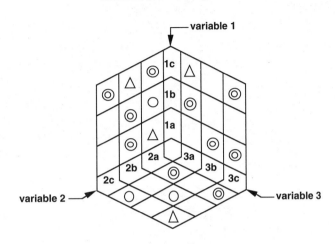

4. X-Shaped Matrix

5. C-Shaped Matrix

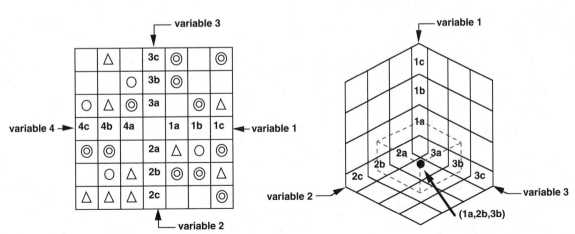

1. L-Shaped Matrix

This is the most basic form of a Matrix Diagram. In the L-Shaped, two interrelated groups of items are presented in line and row format. It is a simple two-dimensional representation that shows the intersection of related pairs of items as shown in Exhibit M-1, which shows the relationship between a company's operations problems and the various maintenance tasks to be done. This exhibit shows that machine lubrication and roll tension have the strongest impact on the greatest number of operations problems. It can be used to display relationships between items in countless operations areas such as administration, manufacturing, personnel, R&D, etc. For example, Exhibit M-2 is used to first identify all of the organizational tasks that need to be accomplished and then how they should be allocated to individuals.* (See page 136 for a complete explanation.)

Exhibit M-1
L-Shaped Matrix
Operations Problems/Maintenance Tasks

1. Maintenance Tasks

	Roll Tension	Pin Setting	Machine Lubrication
Machine Jam	◎	○	◎
Twisted Material	◎	◎	
Soiled Goods			◎

2. Operations Problems

---

* This process is most revealing if the team identifies all of the organization's tasks first and then each person completes the matrix individually. The coding of each task/person responsible can then be compared within the group. You'll be amazed at the differences in perception and the real "loser" tasks for which no one wants to be responsible but everyone wants to be "kept informed."

Exhibit M-2
L-Shaped Matrix
Allocation of Organization Tasks

◎ Primary Responsibility
○ Secondary Responsibility +Slightly More Emphasis
△ Communications/Needs to Know

	Bob	Mike	Lee	Larry	Anna	Jetty	Dona	Brd.Dir.	Other
Administration									
Payroll	◎		△					○	
Benefits	○	△	◎	△	△	△	△	○	
Office Systems	○	○	◎			◎	△		
Computer Programs	○	△	◎			○	○		
Courses									
Update Mailing List			○			◎	◎		
Select Courses to be Offered	◎	◎	◎			△	△		
Approve Course Content	◎	◎	○			△			
Prepare Brochures	○	○	○		◎	△			
Prepare Mailing			△			◎	○		
Hotel Arrangements	△	△	◎			△+	△		
Order Materials	△	△	◎			○	△		
Register People			△		△	◎	○		
Copy Materials					△	◎	○		
Prepare Packets	△	△	△			◎	○		
Room Set-up	◎	◎	◎						
Post Receipts			◎						
Prepare Bills	△		◎				○		
New Course Development									
Market Research	○	◎	△			△			
Implementing Deming	◎	◎	△	○		△	○		
TQC	◎	○	△						
Fundraising									
Annual Reports	○	○	◎		◎	○	△	△	
Corporate Donations	◎	○	○	◎	○				
Committees									
Program Planning	◎	○	△			△			
Statistical Resources	○	◎	△	△	△	△			
TQC	◎	○	△	△	△	△			

Exhibit M-3 shows yet another application to an all-too-common problem: understanding the "voice of the customer." This partial A-1/House of Quality chart allows you to prioritize the customer demands as well as the key elements of the internal system that must be controlled or improved.

Exhibit M-3 reveals two important pieces of information:

1. **Key Customer Demands**
   - Low Starting Effort
   - Start After Storage
   - Doesn't Stall
   - Doesn't Eject Projectiles

2. **Key Substitute Quality Characteristics**
   - Ignition
   - Blade Geometry
   - Tensile Strength of Shear Pin

Therefore, the design process would focus its scarce resources on product/process changes in these areas.

Exhibit M-3
L-Shaped Matrix
QFD A-1 Chart/Lawn Mower Customer Demands

	Starting torque	Power source weight	Fuel consumption	Ignition	Wheel diameter	Wheel bearing type	Blade geometry	Standard parts	Tensile strength of shear pin	Fuel to air ratio	Rate of importance	Company now	Competitor X	Competitor Y	Competitor Z	Plan	Ratio of Improvement	Sales point	Absolute weight	Demanded weight
Highly reliable				39.6			13.2	39.6			5	4	5	3	2	5	1.25	O	7.50	4.4
Low fuel consumption		3.6	10.8	3.6			1.2		1.2	10.8	2	3	3	3	3	3	1.00		2.00	1.2
Low starting effort	65.7			65.7			21.9		21.9	21.9	5	3	5	4	3	5	1.66	◎	12.45	7.3
Simple starting mech.	8.4			8.4					25.2	2.8	3	3	5	3	3	4	1.33	O	4.78	2.8
Starts after storage	5.5			49.5					49.5	16.5	5	4	4	3	2	5	1.25	◎	9.38	5.5
Cuts under all operating cond's.				2.3			20.7		20.7		4	4	4	3	3	4	1.00		4.00	2.3
Doesn't stall			5.5				49.5		16.5	49.5	5	4	5	4	3	5	1.25	◎	9.38	5.5
Can't be started by child	4.4			39.6							4	4	5	4	3	5	1.25	◎	7.50	4.4
Doesn't eject projectiles							16.5				5	4	4	3	3	5	1.25	◎	9.38	5.5
Stops when hits obstructions				3.9			3.9			3.9	3	4	3	3	3	3	.75		2.25	1.3

Total 170.80

	Starting torque	Power source weight	Fuel consumption	Ignition	Wheel diameter	Wheel bearing type	Blade geometry	Standard parts	Tensile strength	Fuel to air
Total	111.9	65.38	28.9	289.81	54.8	41.0	234.6	68.09	227.0	130.0
%	4.09	2.39	1.06	10.59	2.0	1.5	8.57	2.48	8.28	4.75
Company now										
Competitor X										
Competitor Y										
Plan										

= 2816.13 (Total)

◎ = 9
O = 3
△ = 1

When calculating the sales point, the following values are substituted:

◎ = 1.5
O = 1.2
blank = 1.0

*This is a partial example of an A-1 matrix. The totals refelct the sum of all the numbers, some of which are not shown here.

Exhibit M-4 compares known customer demands with brainstormed design concepts. This is loosely based on the New Concept Selection[1] technique developed by Professor Stuart Pugh of the University of Strathclyde in Glasgow, Scotland. This chart simply makes a meets/does not meet evaluation of each design option, as compared to each customer demand. This particular format, however, doesn't factor in the relative importance of each customer demand. A simple alternative would be to use a three level rating scale in which all of the base demands would have one symbol (+ or -), the next higher level would use two symbols (+ + or - -), and the highest level of demand would use three symbols (+ + + or - - -). The cumulative score would simply add up the total number of +'s and -'s , but this weighting would be built in.

[1] Stuart Pugh, *Concept Selection—A Method That Works*, International Conference on Engineering Design—ICED '81, Rome, Italy: March 9-13, 1981.

Exhibit M-4
L-Shaped Matrix
Comparison of Customer Requirements and
Child Safety System Design Options

+ Meets Requirement / — Does Not Meet Requirement		Heated Waterbed System	Adjustable Inflatable System	(Assume seat) Air Bags	Playroom w/Air Bags	Playroom w/Padding	Car Seat w/ Modified Leg Support	Car Seat w/ Reclining	Car Seat w/ Inflatable Padding	Car Seat w/ Sound System	Car Seat w/ Storage (Child Access)	Modular Car Seat by Component	Impact Actvd. Adj. Harness System
Age of Child 0 - 1	Needs Head Support	+	+	—	—	—	+	+	+	+	+	+	
	Needs Soft Padding	+	+	—	+	+	+	+	+	+	+	+	—
	Back Should Absorb Shock	+	+	—	—	—	+	+	+	+	+	+	—
Age 3	Needs Attention Occupiers	—	—		+	+	—	+	—	+	+	+	
	Child Proof Restraint	—	—	+	+	+	+	+	+	+	+	+	+
Stds.	Emergency Release	—	+	+	+	+	+	+	+	+	+	+	+
	Flame Retardant	+	+	+	+	+	+	+	+	+	+	+	+
	Withstand Klb/in2	+	+	+	+	+	+	+	+	+	+	+	+
	Immobile During Use	+	+	+	+	+	+	+	+	+	+	—	+
	Not Toxic	+	+	+	+	+	+	+	+	+	+	+	+
Comfort	Mobility of Head	+	+	+	+	+	+	+	+	+	+	+	+
	Mobility of Arms	+	+	+	+	+	+	+	+	+	+	+	+
	Legs Supported	+	+	—	+	+	+	+	+	+	+	+	—
	Surface Cool in Summer & Warm in Winter	+	+		+	+	+	+	+	+	+	+	
	Play Surface	—	—		+	+	—	—	—	—	—	+	
Adjustable	Grow With Child	—	+	+	+	+	—	—	+	+	—	+	+
	Fits All Cars	+	+	+	—	—	+	+	+	+	+	+	—
	Plusses (+)	12	14	10	14	14	14	15	15	16	15	16	9
	Minuses (-)	5	3	7	3	3	3	2	2	1	2	1	8

2. T-Shaped Matrix

Because a matrix is only two-dimensional (except for the three dimensional C-Shaped Matrix), it can only show relationships between two items at a time. This is often sufficient, but sometimes a user wants to see a third set of items that would provide a more complete implementation picture. The T-Shaped Matrix doesn't create a third dimension but it does provide an additional "leg" that allows for a relationship analysis among three sets of items on the same page. The T-Shaped Matrix still only allows you to compare two sets of items at a time. The third set of relationships can only be inferred and not shown directly.

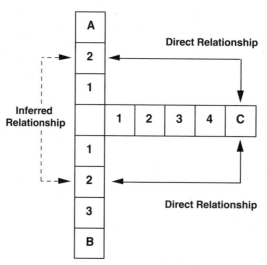

For example, Exhibit M-5 shows the relationship between Operations Problems, Maintenance Tasks, and Products run on this equipment. It allows you to see which operations problems are most closely related to which maintenance tasks (possibly indicating a cause and effect relationship between a maintenance failure and process problem). But the T-Matrix also shows how sensitive each of the products is to those same maintenance failures.

Exhibit M-5
T-Shaped Matrix
Operations Problems/Maintenance Tasks/Products

Exhibit M-6 provides a snapshot of the entire company-wide Continuous Improvement training program. In one chart it actually answers all of the following questions through the relationships shown and the coding applied:

- Who receives training in each topic?
- At what level of detail should the training be supplied?
- Who is responsible for developing each training course?
- Who has individual training responsibility in each course?
- Who is a member of a training team for each topic?

What it doesn't tell us is which trainer interfaces with which trainees. In order to show that type of interaction, you would need to reorient this same input into the next matrix to be reviewed: the Y-Shaped Matrix.

3. Y-Shaped Matrix

The Y-Shaped Matrix allows the user to combine and compare three sets of items (two at a time). For example, Exhibit M-7 (see page 150) reformats the sets of items compared in Exhibit M-5. Notice that the format allows you to now relate the Products (Item Set 3) directly with the Operations Problems.

A knowledgeable manager would see from this that there is a strong similarity between the problems of Products A and C. Notice that this similarity is also present in the portion of the matrix that relates Products with Maintenance Tasks. You could therefore theorize that both products might have fewer "machine jams" and "soiled goods" problems if "roll tension" and "machine lubrication" were improved.

Exhibit M-6
T-Shaped Matrix
Company-Wide Continuous Improvement
Training Program

Who Trains

- Human Resource Dept.
- Managers
- Operators
- Consultants
- Production Operator
- Craft Foremen
- GLSPC Coordinator
- Plant SPC Coordinator
- University
- Technology Specialists
- Engineers

* Need to Tailor to Groups

Courses

SQC | 7 Basic QC Tools | 7 Tools For Mgmt. & Plng. | Reliability | Standardization | QC Basics | QCC Facilitator | Self Inspection | Problem Solving | Communication Skills | Management By Plng. (MBP) | Design of Experiments (DOE) | Total Productive Maintenance | Just In Time (JIT) | New Superv. Training | Total Quality Control (TQC) | Group Dynamics Skills | Quality Function Deploy. (QFD)

Who Attends

- Executives
- Top Management
- Middle Management
- Prod. Supervisors
- Supp. Func. Mgrs.
- Staff
- Marketing
- Sales
- Engineers
- Clerical
- Prod. Worker
- Qual. Professional
- Project Team
- Emp. Involv. Teams
- Suppliers
- Maintenance

Key for "Who Trains"

⊙ Development Responsibility
○ Individual Training Responsibility
△ Team Member

Key for "Who Attends"

X = Full

O = Overview

Exhibit M-7
Y-Shaped Matrix
Operations Problems/Maintenance Tasks/Products

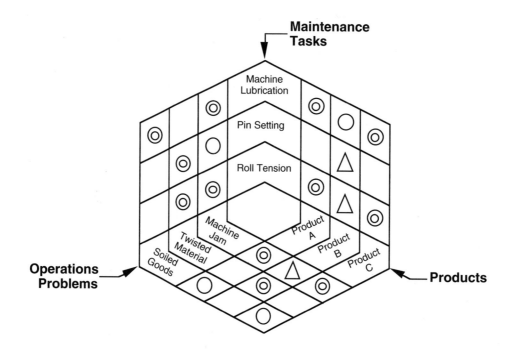

4. X-Shaped Matrix

The X-Shaped Matrix lets you now show four <u>sets</u> of relationships in one document. It does this by adding one more axis to the T-Shaped Matrix. It still only allows you to relate two Item Sets at a time.

Exhibit M-8 (see page 152) now adds the dimension of the shift variable as a possible contributor to the operations and maintenance problems.

There are several notable things about Exhibit M-8. First, it does not use the traditional relationship symbols. Instead, it uses the relevant percent in each relationship. For example, the upper left quadrant shows the percentage of each product's total production that is needed on shift. The lower left quadrant represents the breakdown of operations problems over three shifts, e.g., 45% of all machine jams occur on the third shift; 45% on the second shift; 10% on the first shift.

Conclusions based on Exhibit M-8:

- Machine jams is the operations problem most affected by maintenance (or lack of it).

- Roll tension and machine lubrication are the most critical maintenance tasks to perform.

- Roll tension and machine lubrication are maintenance problems for Products A and C, but not Product B.

- Most of Products A and C are produced on the second and third shifts.

- Second and third shifts do have most of the operations problems.

Exhibit M-8
X-Shaped Matrix
Operations Problems/Maintenance Tasks/
Products/Shifts

3. Products

				Roll Tension	Pin Setting	Machine Lubrica-tion	
40%	35%	25%	Product C	◎		◎	
	20%	80%	Product B	△	△	○	
46%	39%	15%	Product A	◎		◎	
Third Shift	Second Shift	First Shift		Roll Tension	Pin Setting	Machine Lubrica-tion	**1. Maintenance Tasks**
45%	45%	10%	Machine Jams	◎	○	◎	
38%	32%	30%	Twisted Material	◎	◎		
41%	48%	11%	Soiled Goods			◎	

4. Shifts

2. Operations Problems

- **THEREFORE**
  - a) Find out if the first shift really does maintenance that much better than the second and third shifts,

OR

  - b) Find out whether Product B is that much easier to run than Products A and C that it would account for so few operations problems. Remember that the first shift produced almost all of Product B.

5. C-Shaped Matrix

The C-Shaped Matrix (or Cubic Type Matrix) makes it possible to visually represent the relationships among three sets of items simultaneously. In theory, this seems like an interesting method. However, it is rarely used. Why?

1) It is difficult to draw without computer software.

2) Users often feel uneasy about using subjective judgment to make something that begs for objective measurement, e.g., Designed Experiments, Taguchi Methods.

3) It can be confusing for the "layman" to interpret.

Below is a portion of Exhibit M-7 (Y-Shaped Matrix) displayed as a three dimensional C-Shaped Matrix. Exhibit M-9 reveals two sets of three-way relationships:

✳ 1) Product A has an unusually high incidence of machine jams; these jams are most closely related to the roll tension maintenance.

● 2) Product C has a serious soiled goods problem that can be controlled by improved machine lubrication procedures.

Another possible use of a C-Shaped Matrix is to display the results of Designed Experiments (either Full Factorial or Taguchi Methods). It could show the optimum combinations of levels within the factors chosen. Of course, this would be limited to an experiment with three or less factors or to display the factors shown to be the most influential.

Exhibit M-9
C-Shaped Matrix
Operations Problems/Maintenance Tasks/Products

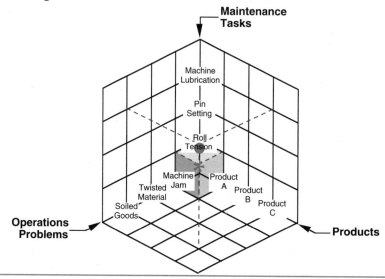

6. Tree Matrix

One of the most commonly used variations on the Matrix Diagram is a format that combines either one or two Tree Diagrams into an L-Shaped Matrix. This form takes advantage of the fact that the Tree Diagram explodes very vaguely defined issues or tasks into specific items to be completed. By combining the two forms you can:

- Have confidence that you have generated the most complete set of items that should be practically considered.

- Uncover the unknown relationships between the two very complete lists of items.

For example, in the A-1 Chart of QFD design system, both the Customer Demands and the Substitute Quality Characteristics are the last level of detail in Tree Diagrams. The most common format is to have a Tree Diagram form the vertical side of an L-Shaped Matrix. The horizontal side of the matrix consists of a known set of criteria or considerations. For example, the following two matrices show how both formats can be used to plan improvements in maintenance procedures.

Exhibit M-10 shows that there are at least five instances in both implementation plans where the success of one plan would have a strong impact on the success of the other. For example, if the plan under roll tension was to make operator adjustment impossible, this might be moving in the opposite direction of operators doing lubrications themselves. One suggests "hands-on involvement" the other "hands-off" monitoring.

Exhibit M-10
Tree Matrix (Double)
Roll Tension/Machine Lubrication

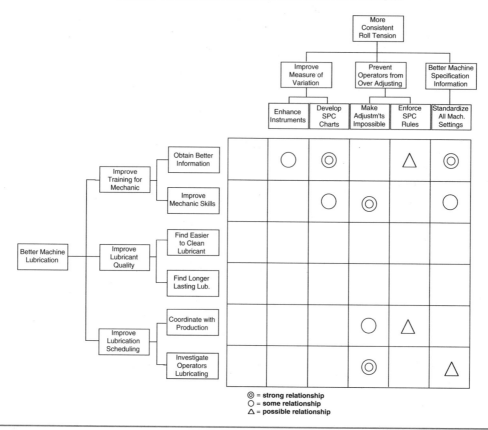

⊙ = **strong relationship**
○ = **some relationship**
△ = **possible relationship**

Exhibit M-11 shows clearly that improving machine lubrication is not a maintenance task. In fact, maintenance doesn't have primary responsibility for any of the tasks. They are a member of the team. Notice that this also allows you to assess the role of untraditional "players" such as union leaders. Often, they get involved only after important decisions have been made and chaos follows.

Exhibit M-11
Tree Matrix (Single)
Machine Lubrication/Implementation Responsibility

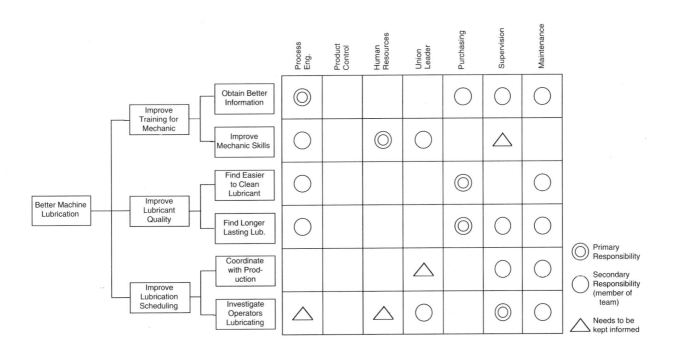

Construction of a Matrix Diagram

The process of constructing any of the various format Matrix Diagrams is very straightforward. Your understanding of which matrix format will shed the most light on your problem is the most critical ingredient. The steps, along with illustrations from the Frankel Corporation, are described below.

 Select Key Factors Affecting a Successful Implementation

Every time we generate a list of options or actions to be taken we must also decide what is going to make or break any implementation plan developed. For example, some critical considerations might be:

- Who is involved in execution either as the implementation team leader or team member?

- Who's <u>not</u> involved in implementation?

- Prioritize actions based on criteria such as cost, speed of implementation, etc.

- Relate proposed actions to current programs to avoid conflicts.

A **The IMP System Marketing Team** that developed the Tree Diagram and the Prioritization Matrices decided to next define who would have implementation responsibility for each of the most important tasks identified from these diagrams/tables. In addition, the team wanted to use the Matrix Diagram to identify related and sequential tasks within the same list.

B **The Missed Promised Delivery Dates Team**, based on the conclusions from the Prioritization Matrices in Chapter Four, decided to focus on the tasks involved in the "reduction of data entry complexity." The team felt they needed to allocate two things:

- Implementation responsibility across departments/functions

- The resources needed to implement

2 Assemble the Right Team

It may seem like a cliché, but it really is true that any analysis is only as good as the team assembled to complete the study. In Matrix analysis, once the tasks and the sets of items that they will be related to are chosen, then the correct team must be assembled to make those relationship decisions. In the case of a responsibility matrix, those individuals likely to be assigned tasks should be involved in either the original assignment of responsibility or a review of assignments done by a smaller team.

Select the Appropriate Matrix Format

At this point the questions could not be any simpler.

a) How many key considerations in implementation have you identified in Step #1?

b) Given that number, which matrix format would give you the greatest insight into the implementation plan?

Let's review the possible matrix formats:

- L-Shaped: 2 sets of items.

- T-Shaped: 3 sets of items showing both indirect and direct relationships.

- Y-Shaped: 3 sets of items showing direct relationships.

- X-Shaped: 4 sets of items showing both indirect and direct relationships.

- C-Shaped: 3 sets of items showing simultaneous relationships.

3 Select The Appropriate Matrix Format (Cont'd.)

NOTE: <u>How Critical Is the Choice of Format?</u> The most important decision is not how you structure the matrix, but which specific items you choose to compare for interrelationships. There is no science to this. The team's experience suggests that certain items might be related. The team selects those items, puts them into a matrix format that seems helpful, and then begins to test the original premise, which might or might not be supported.

A **The IMP System Marketing Team**, which is focusing on "Using The Best Mix of Marketing Mediums," chose to use an L-Shaped Matrix to show the allocations of implementation responsibilities and the sequencing of tasks. (See page 162.)

B **The Missed Promised Delivery Dates Team** opted for a T-Shaped Matrix to show implementation responsibility and the resources required to implement. (See page 163.)

A IMP System Marketing Team
L-Shaped Matrix
Using the Best Mix of Marketing Mediums

◎ Primary Responsibility ○ Secondary Responsibility △ Kept Informed	Marketing	Sales	Human Resources	Safety Engin.	Legal	Quality Assurance	R & D	Sequencing of Tasks
Develop Product Tests Set in Both the Laboratory and Everyday Situations								
Obtain Federal Product Requirement Standards								
Hire Marketing Manager with Children's Product Experience								
Mix Visual, Auditory, and Hands-On Record of IMP System Performance								
All Photographs Showing Side-By-Side Testing of Comparable Products								
Develop Ease-of-Use Oriented Ads								
Develop Product Demonstration That Is Portable, Repeatable, Dramatic and Foolproof								
Outline Convincing Sales Points for Dealer Network								

B Missed Promised Delivery Dates Team
T-Shaped Matrix
Reducing Data Entry Complexity

Legend:
- ◎ High
- ○ Medium
- △ Low

- ◎ 9
- ○ 3
- △ 1

- ◎ Prime Responsibility
- ○ Secondary Responsibility
- △ Kept Informed

	Increase Monitor Size	Menu Driven	Improve Prompts	Increase Size of Type	Color Code Forms by Prod. Groups	Shorten 11-Digit Prod. Code	More Obvious Diff. Among Prod Grp.Codes	
Resources Needed								Total
Capital Investment								
Staff Time								
Training Time								
Space								
Equipment Availability								
People								
Purchasing								
Software								
Hardware								
MIS								
HRD								
Distribution								
Production								
								Total

Choose and Define Relationship Symbols

There is literally no limit to the variety that can be used to indicate the kind of relationship that exists between any two items in a matrix. In fact, a matrix user can develop any graphical symbol as long as there is one thing provided: **A legend.** A legend should define the meaning of each symbol so clearly that interpretation is consistent even if someone was not involved in the original process.

The most common symbols fall into two categories:

A. Strength of Relationship

◎ = Strong Relationship
○ = Some Relationship
△ = Weak/Possible Relationship

A subset of this category is **Supportive vs. Negative Relationships**

◎ + - Strong Supportive/Negative
○ + - Some Supportive/Negative
△ + - Weak or Possible Supportive/Negative

Another group in this category refers to the direction of the relationship. This is simply indicated by arrows accompanying the strength symbols.

Choose and Define Relationship Symbols (Cont'd.)

	4	5	6
1	↑◎	←△	
2		←◎	↑△
3	↑◎	↑△	

In this example, it shows that if you work on Item #1, it will have a strong effect on Item #4, whereas Item #5 would have only a possible or weak effect on Item #1.

NOTE: <u>Can You Have Two-Way Arrows?</u> Many (if not most) relationships in fact do have some cause and effect impact going both ways. However, it must be decided in the Matrix Diagram which is the <u>strongest</u> influence. It is too easy to simply draw a two-headed arrow. The real insight comes when the team comes to a consensus around the direction of the arrow.

B. Level of Responsibility

◎ = Primary Responsibility
○ = Secondary Responsibility
△ = Should be kept informed/May need information from them

Choose and Define Relationship Symbols (Cont'd.)

NOTE: <u>Can there be shared primary responsibility</u>? Whenever a responsibility matrix is used it is tempting to split primary responsibility! It has been our experience that shared primary responsibility equals no accountability. The secondary responsibility symbol allows for designation of team members.

NOTE: <u>Symbols or Numbers</u>? It is common to associate a numerical value with the various symbols already mentioned. This allows the user to get some composite numbers that are helpful for prioritizing tasks, etc. If this is done, why not just use numbers? The Japanese, who brought the Seven Management and Planning Tools together, are a highly visual people. This allows managers in many different facets of the operation to quickly assess the status of any situation. Likewise, the symbols encourage "emergent" thinking so that you allow the tool to surface patterns that are easy to use graphically. These patterns are often buried in numbers.

Complete the Matrix

Assuming:

. . . The right tasks to be implemented

. . . The right items relate to each other

. . . The appropriate matrix format for the task at hand

. . . Well chosen and clearly defined relationship symbols

DO THE MATRIX ! ! !

A **The IMP System Marketing Team** completed their L-Shaped Matrix on the Best Mix of Marketing Mediums. It defined both the allocation of responsibility and the sequence of those key tasks. (See page 168.)

B **The Missed Promised Delivery Dates Team** completed their T-Shaped Matrix that matrixed the key tasks in Simplifying Order Entry with those functions responsible and needed resources. (See page 169.)

A IMP System Marketing Team
Completed L-Shaped Matrix
Best Mix of Marketing Mediums

	Marketing	Sales	Human Resources	Safety Engin.	Legal	Quality Assurance	R & D	Sequencing of Tasks **
◎ Primary Responsibility / ○ Secondary Responsibility / △ Kept Informed								
Develop Product Tests Set in Both the Laboratory and Everyday Situations	○	△	△	○	△	◎	○	2.1
Obtain Federal Product Requirement Standards	△	△		◎	○	○	△	2.0
Hire Marketing Manager with Children's Product Experience	◎	○	○					1.0
Mix Visual, Auditory, and Hands-On Record of IMP System Performance	◎	○		○	△	○	○	2.2
All Photographs Showing Side-By-Side Testing of Comparable Products	◎	○		○	○	○	○	3.2
Develop Ease-of-Use Oriented Ads *	◎	○		△	△		○	3.1
Develop Product Demonstration That Is Portable, Repeatable, Dramatic and Foolproof *	◎	○	△	○	△	△	○	3.0
Outline Convincing Sales Points for Dealer Network +	○	◎	○	△	△			4.0

* Referring back to Exhibit P-7, on page 133, notice that these two tasks have about an even split on "in" and "out" arrows, but the overall strength of relationship is very strong as well as a high total number of arrows.

+ This task actually has the highest strength relationship in the entire list. However, nearly all (11 of 14) of the arrows lead to (not come from) this item. It is therefore not a basic cause, but the key outcome to strive for in the entire process. It becomes the measure of success. In this case, the process will be a success if all of the tasks convince the dealer network to push the product to customers as part of the options package.

** These numbers indicate both the sequence and relationships among tasks. For example, the task "Hire Marketing Manager . . ." would be the first to be done. Therefore it's marked by a 1.0. The second task would be to "Obtain Federal Product Requirement Standards as marked by a 2.0, etc. All of the other tasks beginning with a 2 (e.g., 2.1, 2.2) are related, but follow in sequence. All of the 3's are related and sequenced as well.

Notice that this is not a traditional matrix column in which you place a relationship symbol. This is more like a typical table column with values to be filled in.

B Missed Promised Delivery Dates Team
Completed T-Shaped Matrix
Simplifying Order Entry

Legend:
- ◎ High
- ○ Medium
- △ Low

- ◎ 9
- ○ 3
- △ 1

- ◎ Prime Responsibility
- ○ Secondary Responsibility
- △ Kept Informed

Resources Needed	Increase Monitor Size	Menu Driven	Improve Prompts	Increase Size of Type	Color Code Forms by Prod. Groups	Shorten 11-Digit Prod. Code	More Obvious Diff. Among Prod Grp.Codes	Total
	21	19	13	10	9	21	15	Total
Capital Investment	◎	△	△	△	○	○	○	21
Staff Time		◎	◎	◎	○	◎	◎	48
Training Time		◎	○		○	◎	○	27
Space	○							3
Equipment Availability	◎							9

People / Purchasing	Increase Monitor Size	Menu Driven	Improve Prompts	Increase Size of Type	Color Code Forms by Prod. Groups	Shorten 11-Digit Prod. Code	More Obvious Diff. Among Prod Grp.Codes	Total
Purchasing	○				○			6
Software	○	◎	◎	◎		○	○	36
Hardware	◎	○	○	○				18
MIS	○	○	○	○		◎	◎	30
HRD					△			1
Distribution						△	△	2
Production	△	△	○	△	◎	○	○	21
								Total

Summary

The matrices described in this chapter are clearly variations on a theme. It suggests that a team of knowledgeable people can decide whether any two items are related. The cumulative effects of these one-to-one relationships are the **patterns** that can truly surprise us. This requires us to trust the process and reserve judgment until the matrix is completed. We now know the priority tasks and who is responsible for implementing them. Therefore, these identified responsible individuals/functions must do the time and contingency planning portion of the process. These will be covered in the following two chapters: Process Decision Program Chart (PDPC) and the Activity Network Diagram.

Other Sources of Information

There are other resources that can help you learn about the uses and techniques of constructing the Matrix, or how to *facilitate* the use of the Matrix, as well as other management and planning tools. Several of these resources are:

- *The Memory Jogger*™ *II*
- *Coach's Guide to The Memory Jogger*™ *II*
- *The Coach's Guide* Package
- *The Learner's Reference Guide to The Memory Jogger*™ *II*
- *The Educators' Companion to The Memory Jogger Plus+*®
- *The Memory Jogger Plus+*® Software
- *The Memory Jogger Plus+*® Videotape Series

Chapter 6

Process Decision Program Chart (PDPC)

Definition

The Process Decision Program Chart (PDPC) is a method that maps out conceivable events and contingencies which can occur in any implementation plan. It in turn identifies feasible countermeasures in response to these problems. This tool is used to plan each possible chain of events that needs to occur when the problem or goal is an unfamiliar one.

The underlying principle behind the PDPC is that the path toward virtually any goal is filled with uncertainty and an imperfect environment. If this weren't true, we would have a Deming "sequence" like the following:

Plan⎯⎯⎯⎯→Do

Reality makes the Deming Cycle a necessity.

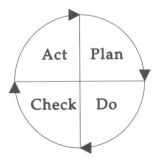

The PDPC anticipates the unexpected and, in a sense, attempts to "short circuit" the cycle so that the "check" takes place during a dry-run of the process. The beauty of the PDPC is that it not only tries to anticipate deviations, but also to develop countermeasures that will either:

a) prevent the deviation from occurring, or

b) be in place in case the deviation occurs

The first option is ideal in that it is truly preventive. However, we live in a world of limited resources. In allocating these resources we have to often "play the odds" as to the chance of X, Y, or Z happening. Given that fact, the next best thing is to have a contingency plan in place when a case occurs that we were "betting against." The PDPC provides a structure to go in either direction.

When to Use a PDPC

In the simplest terms possible, a PDPC is used whenever uncertainty exists in a proposed implementation plan. Of course, this would include virtually any plan ever written. The keys to remember in properly selecting a PDPC as a tool are:

1. The task at hand should be one that is either new or unique. A task that is routine normally doesn't warrant a PDPC unless a major new factor is introduced, such as a major market or personnel change.

2. The implementation plan should have sufficient complexity. If the steps are so few or so clear that deviations are trivial or self-explanatory, then a PDPC would be a wasted effort.

3. The stakes of potential failure should be high.

4. The efficiency of the implementation must be critical. If, for instance, there is a 12-month window within which a 3-month plan must be implemented, there is plenty of "slack time" for deviations from the original path.

5. The contingencies must be plausible. We must be creative, but cautious that we don't create problems where none could possibly exist.

Broad Uses of a PDPC

There are two major use patterns of a PDPC. Both are related to when you use the tool in the planning cycle. The one that has been most widely used in the United States is the Pre-Planned PDPC. Simply stated, this type of PDPC is used to anticipate any problems before doing anything toward implementation. The different paths and possibilities are brainstormed and are only limited by the time, knowledge, and/or imagination of the user. (See page 174, Exhibit PD-1, Planning An Exotic Family Vacation.)

This Pre-Planning use is usually an exercise in Contingency Planning in answer to the question "What could be done as a countermeasure if 'A' happened? If 'B' happened?" and so on. This is typically used in applications such as safety and large scale program planning. In both these cases, it is too costly or dangerous to move out of the planning and into the doing phase. These usually are based on analyzing each implementation step in sequence. When added together these steps lead to an end goal that is to be achieved.

Exhibit PD-1
PDPC
Planning an Exotic Family Vacation

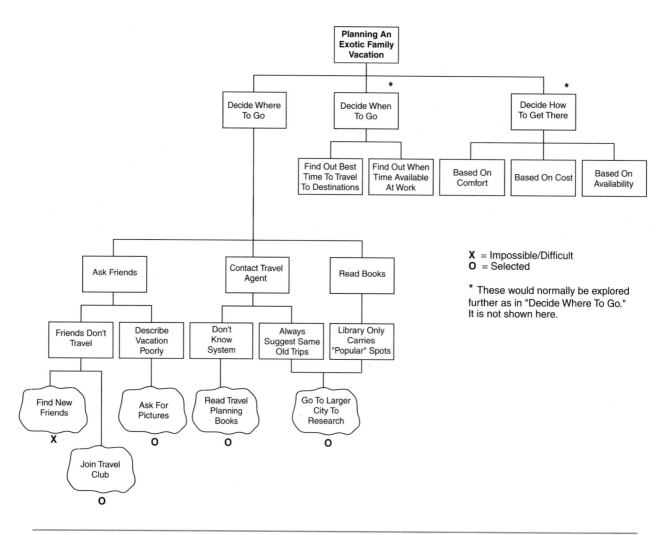

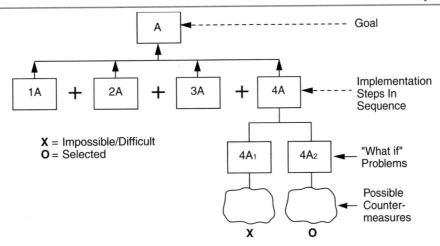

The Pre-Planned PDPC can also be applied when <u>Alternative Implementation Plans</u> are being evaluated.

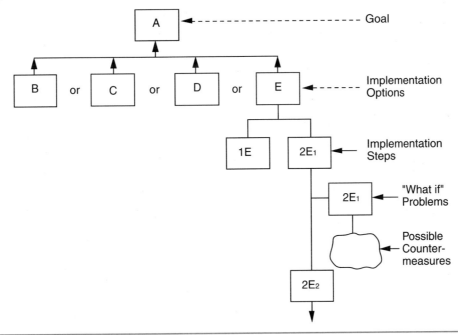

Common sense dictates that the implementation path with the fewest or least serious "what if's" is the best alternative.

An alternative (less widely used) is the <u>Real Time PDPC</u>. In this format it is used to test theories and alternatives as they are implemented. (See Exhibit PD-2 on page 177.)

In this application, the initial potential problems are brainstormed, but where the process goes is dependent on how the alternatives develop. Therefore, rather than evaluating and selecting countermeasures up front (Pre-Planned PDPC), they are only selected once they have been tried and evaluated.

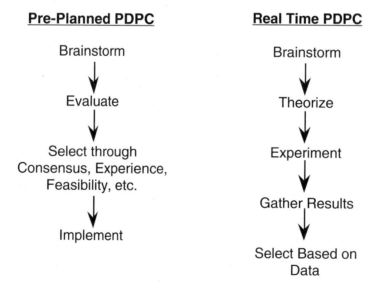

Exhibit PD-2
PDPC
Improve Fibre Retention

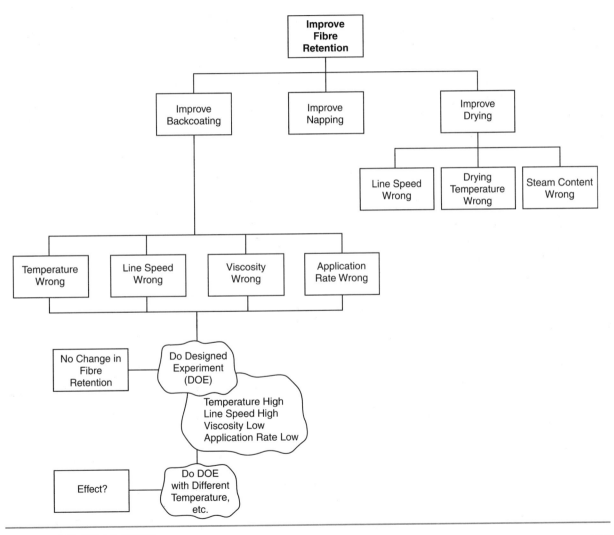

Typical Uses of a PDPC

- An organization is planning its annual conference with a special focus on registering attendees smoothly and efficiently. The stakes are high (upset attendees), there are major changes from the past (new location, double the number of attendees), and the process must be efficient since 800 people have to register in one hour.

- A manufacturer is planning a production equipment installation during its summer shutdown. The stakes are high (no production capacity if the installation falls behind schedule), efficiency is critical (a 7 day shutdown with a 6-1/2 day project) and the task is unique (the first computer controlled line).

- An insurance company is planning to write liability insurance in a new industry (new for the insurance company and as an identifiable market). It is potentially lucrative but with unknown risks. Executives might use a PDPC because the stakes are very high (unknown exposure, high profits, market domination), timing is critical (excessive delays could give other companies the chance to enter the market), and the topic complex (licensing in 12 states and 3 foreign countries).

- A Director of Human Resources is designing the first ever gain-sharing program in her company. A PDPC is in order since this is certainly a new effort (the closest thing in the past was a $100 bond to the person with the best suggestion for the year), the stakes are high (mess with anything but not my pay check) and it is complex (gaining management and employee support, implementing suggestions, creating and maintaining an equitable formula, etc.).

All of these uses ensure one thing: unexpected twists and alternatives will happen in any implementation plan. The key is to anticipate and plan for them in the conceptual stage. It is faster, cheaper, and more effective. The PDPC brings the disciplined logic seen in the previous tools to bear on the imperfect plan.

Construction of a Process Decision Program Chart (PDPC)

Even though the PDPC is a methodical process, it has few guidelines in terms of the process and finished product. Unlike the other tools, which have a distinctive final appearance, the PDPC could produce two examples that look radically different. Regardless of the specific format created, it must <u>clearly</u> indicate expected deviations and possible countermeasures.

> ### **1** Assemble the Right Team
>
> As you get closer to implementation, the need to get "in the trenches" input increases. This is even more critical for the PDPC because "what if's" require intense knowledge of the system and its performance gaps. These are often not known or acknowledged higher in the organization. The team must, therefore, have two components:
>
> a) Managers/employees with an overview of the entire process flow, who can describe the sequence of events in any plan.
>
> b) Managers/employees who are experts in each step in the process who can describe problems that are likely to occur.

 The IMP System Marketing Team decided that the first two tasks in their implementation plan (see page 168, Chapter 5, Matrix Diagram), "Hire marketing manager with children's product experience" and "Obtain federal product requirement standards," were fairly straightforward and not appropriate for a PDPC. They simply assigned responsibilities as shown (page 168). However, for the task to "Develop product tests set in both the lab and everyday situations," the team felt that this was unknown territory and appropriate for a PDPC. As shown on page 168, the team was led by QA with strong participation by R&D, Safety Engineering, and Marketing. This group mapped out the basic sequence of events while the other "informed" functions (sales, legal, and H.R.) reviewed the sequence and identified problems at each step.

B **The Missed Promised Delivery Dates Team** felt that all of the tasks in the Responsibility/Resource Matrix (see page 169, Chapter 5) were familiar enough to be assigned to the appropriate functions. They felt that the Activity Network Diagram would be a more helpful tool so they opted not to use the PDPC.

2 Determine the Basic Flow of Proposed Activities

Think of this step as building a skeleton which includes the basic implementation flow. This can be a straight line or a simultaneous flow. (See page 181.)

NOTE: How Detailed in the Process Steps? One of the legitimate problems with the PDPC is simple…Detail EXPLOSION! Therefore, common sense must reign. The implementation steps must stay at a fairly broad level or the user will get lost in a swamp of detail.

② Determine the Basic Flow of Proposed Activities (Cont'd.)

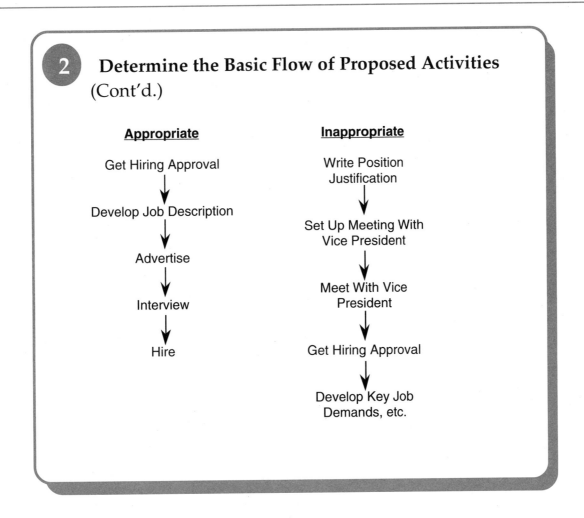

Appropriate	**Inappropriate**
Get Hiring Approval	Write Position Justification
↓	↓
Develop Job Description	Set Up Meeting With Vice President
↓	↓
Advertise	Meet With Vice President
↓	↓
Interview	Get Hiring Approval
↓	↓
Hire	Develop Key Job Demands, etc.

A **The IMP System Marketing Team** developed the following flow of basic steps in the Product Test Development process. (See Exhibit PD-3, page 182.)

Exhibit PD-3
Develop Product Tests Set in Both
the Laboratory and Everyday Situations

Review Federally Mandated Tests

Review Current Lab Capabilities
to Meet Federal Test

Identify Lab Needs for Federal Testing

Identify Capabilities in the Lab
for All Other Testing Needs

Identify Types of Tests That
Impress IMP Target User Group

Identify Gaps Between IMP User Group
Test Needs and Current Lab Capabilities

Develop Visual Examples That
Mirror Lab Tests

Determine Cost of Fitting the Lab Needs and
Reproducing the Results in the "Outside World"

Allocate Budget

Allocate Staff

Choose the Most Workable Chart Format

A tool of any kind is only useful when it is useable. This is especially true with these Tools for Management and Planning. The power of these techniques is in their frequent use. If we can't make them "approachable" then they will be assigned to a specialized planning function and out of the mainstream of day to day managing. These tools are heartily recommended for use by planning specialists, but they already have their "Planning Technique Toolbox" while line and staff managers have been left to their own devices.

Up to this point, all of the tools have been in a graphical format. They have been highly visual. This has tremendous advantages:

1. The user can step back and let new patterns of thought emerge based on patterns, etc.
2. Obvious gaps can be seen and corrected.
3. The information can be easily shared in groups.

The PDPC is the first (and only) technique of the Seven Management and Planning Tools that has a useful non-graphical alternative. Therefore, there are two format choices:

- **Graphical**: The PDPC is a combination of a Tree Diagram and a Process Flowchart.
- **Outline**: The PDPC is a numerically coded running list including steps, problems/contingencies and countermeasures.

Referring back to the example of "Planning an Exotic Family Vacation," below is the same partial PDPC in both the graphical and outline formats:

<u>Graphical Format</u>: Planning an Exotic Family Vacation

Exhibit PD-1
PDPC
Planning an Exotic Family Vacation

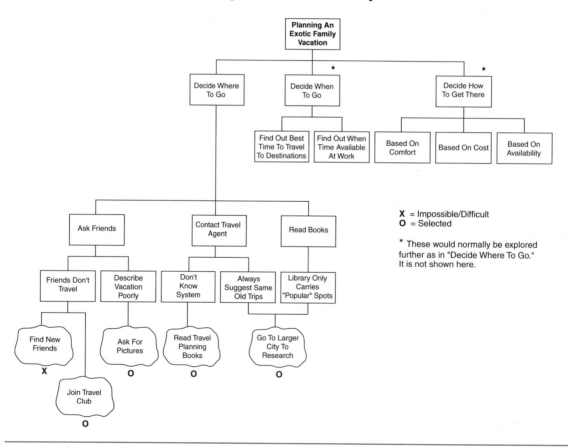

X = Impossible/Difficult
O = Selected

* These would normally be explored further as in "Decide Where To Go." It is not shown here.

Outline Format: Planning an Exotic Family Vacation

Exhibit PD-4
PDPC
Planning an Exotic Family Vacation

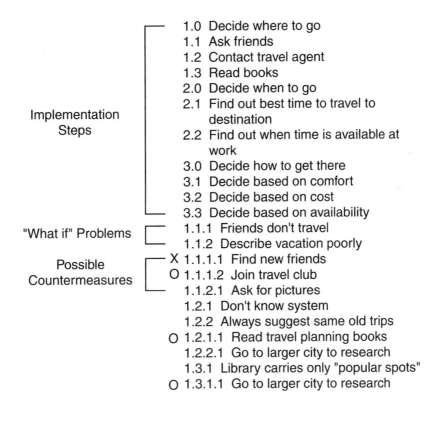

Implementation
Steps

1.0 Decide where to go
1.1 Ask friends
1.2 Contact travel agent
1.3 Read books
2.0 Decide when to go
2.1 Find out best time to travel to destination
2.2 Find out when time is available at work
3.0 Decide how to get there
3.1 Decide based on comfort
3.2 Decide based on cost
3.3 Decide based on availability

"What if" Problems

1.1.1 Friends don't travel
1.1.2 Describe vacation poorly

Possible
Countermeasures

X 1.1.1.1 Find new friends
O 1.1.1.2 Join travel club
1.1.2.1 Ask for pictures
1.2.1 Don't know system
1.2.2 Always suggest same old trips
O 1.2.1.1 Read travel planning books
1.2.2.1 Go to larger city to research
1.3.1 Library carries only "popular spots"
O 1.3.1.1 Go to larger city to research

X = Impossible/difficult
O = Selected

Choose the Most Workable Chart Format (Cont'd.)

NOTE: <u>To List or Not to List</u>? As noted on page 183, there are consistent advantages to using graphical formats whenever possible. Why introduce a non-graphical option at this late stage?

The Outline Format has some distinct advantages:

1. It focuses on the thorough examination of all reasonable contingencies and not the way that they should be drawn.

2. It saves time since no graphical format needs to be drawn and redrawn to maximize clarity.

3. The items are easily traced through the simple outline numbering system. This enables the PDPC to be easily entered into a computer data base for planning and tracking.

4. It can be done as a group or individually. A numbered section can be easily separated and assigned because it doesn't have to fit into any larger graphic. As long as there is consistent numbering, the lists can be easily merged.

3 Choose the Most Workable Chart Format (Cont'd.)

The Outline Format also has some distinct disadvantages:

1. Patterns aren't obvious as in the graphical.

2. It's difficult to use for communicating in a group presentation.

3. It is difficult to show simultaneous paths. Therefore, it tends to be more useable in a simple straight line subject.

On balance, if the user's goal is to make what if/contingency planning a matter of habit for mainstream managers, the Outline Format is the most likely to be used frequently. This alone is often enough to outweigh any of its disadvantages.

A **The IMP System Marketing Team** felt strongly that it had learned to see the power of the graphical methods thus far. The team therefore decided to use the Graphical Format of the PDPC.

Construct the PDPC Using the Chosen Format[†]

As highlighted in Step 3, there are two optional formats: Graphical and Outline. The following construction steps will provide steps for each option.

Graphical Format

<u>Construct A Modified Tree Diagram</u>: The following steps include changes that make the Tree Diagram "modified" from the traditional format.

a) Place Branches in Sequential Order of Implementation: The first level of detail (the first main branches of the Tree) are listed in sequential order of implementation. In the traditional Tree Diagram, this sequencing is shown within branches as the details get more and more specific.

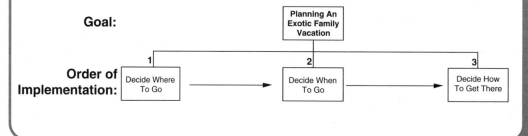

[†] The steps listed will assume the construction of a "Pre-Planned PDPC" rather than a "Real Time PDPC" as described earlier in this chapter.

4 Construct the PDPC Using the Chosen Format (Cont'd.)

b) Go No Further Than the Second Level of Implementation Detail: It is suggested that the Tree Diagram stay at the First Level of detail or at most one more level. Otherwise, "Information Overload" can occur.

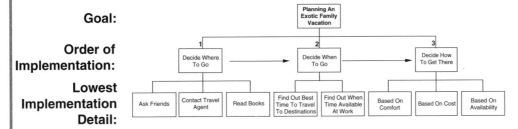

Goal:

Order of Implementation:

Lowest Implementation Detail:

c) Position the Tree Vertically or Horizontally: The Tree Diagram can be placed either vertically or horizontally depending on the number of implementation steps. When there are a high number of steps, it seems to work best with the Tree working from left to right (horizontally). Fewer steps work best in a Tree moving up and down on the page (vertically).

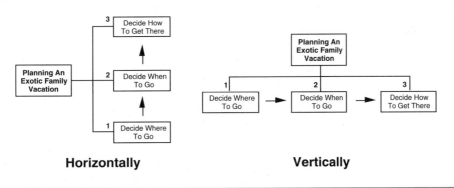

Horizontally   **Vertically**

4 Construct the PDPC Using the Chosen Format (Cont'd.)

d) Ask "What could go wrong?" or "What various/unexpected paths could this step take?" of the lowest level of implementation detail listed.

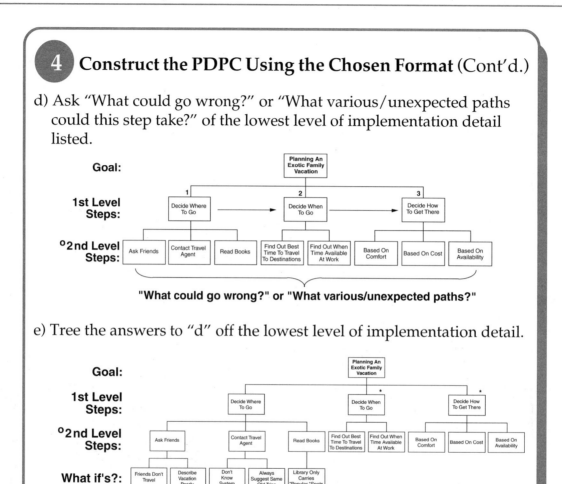

e) Tree the answers to "d" off the lowest level of implementation detail.

°Optional additional detail. It's recommended to stop at the 1st level steps whenever possible.

*These would normally be explored further as in "Decide Where To Go." It is not shown here.

4 Construct the PDPC Using the Chosen Format (Cont'd.)

f) Brainstorm possible countermeasures for each of the contingencies generated in "e" and branch them off each item within cloud-like captions.

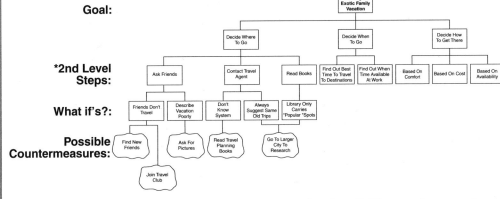

g) Evaluate each possible countermeasure for feasibility and necessity and mark accordingly:

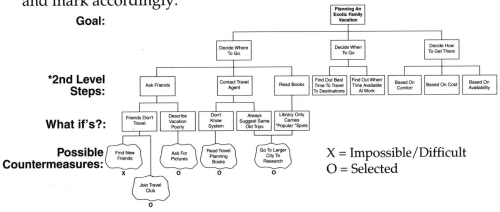

X = Impossible/Difficult
O = Selected

4 **Construct the PDPC Using the Chosen Format** (Cont'd.)

Outline Format

a) List all of the implementation steps broken down to the broadest reasonable level of detail. Indicate the major steps with whole numbers and each substep with a decimal number.

    1.0 Decide where to go
    1.1 Ask friends
    1.2 Contact travel agent
    1.3 Read books
    2.0 Decide when to go
    2.1 Find out best time to travel to destination
    2.2 Find out when time is available at work
    3.0 Decide how to get there
    3.1 Decide based on comfort
    3.2 Decide based on cost
    3.3 Decide based on availability

b) Ask "What could go wrong?" or "What various/unexpected paths could this step take?" of the lowest level of implementation detail listed. This could be either a whole number step (if no decimal substeps) or a decimal substep.

4 **Construct the PDPC Using the Chosen Format** (Cont'd.)

c) Number the "what if" statement to match the substep and place it after the total list of implementation steps. <u>Do only one implementation substep at a time.</u>

1.0	Decide where to go
1.1	Ask friends
1.2	Contact travel agent
1.3	Read books
2.0	Decide when to go
2.1	Find out best time to travel to destination
2.2	Find out when time is available at work
3.0	Decide how to get there
3.1	Decide based on comfort
3.2	Decide based on cost
3.3	Decide based on availability
1.1.1	Friends don't travel
1.1.2	Describe vacation poorly

d) Brainstorm possible countermeasures for each of the contingencies generated in "c" and list them immediately following these items. Number them based on the contingency that they are solving.

1.0	Decide where to go
1.1	Ask friends
1.2	Contact travel agent

4 **Construct the PDPC Using the Chosen Format** (Cont'd.)

1.3	Read books
2.0	Decide when to go
2.1	Find out best time to travel to destination
2.2	Find out when time is available at work
3.0	Decide how to get there
3.1	Decide based on comfort
3.2	Decide based on cost
3.3	Decide based on availability
1.1.1	Friends don't travel
1.1.2	Describe vacation poorly
1.1.1.1	Find new friends
1.1.1.2	Join travel club
1.1.2.1	Ask for pictures
1.2.1	Don't know system
1.2.2	Always suggest same old trips
1.2.1.1	Read travel planning books
1.2.2.1	Go to larger city to research
1.3.1	Library carries only "popular spots"
1.3.1.1	Go to larger city to research

e) Evaluate each possible countermeasure for feasibility and necessity and mark accordingly. As in the Graphical Method:

X = Impossible/Difficult
O = Selected

4 **Construct the PDPC Using the Chosen Format** (Cont'd.)

Our vacation planner evaluated all of the options under "Decide where to go" and made the following judgments:

X 1.1.1.1   Find new friends was not an attractive option.
O 1.1.1.2   Join a travel club was a good idea since it would be a great source for new vacation ideas and meeting new people with similar interests.
O 1.2.2.1   Go to larger city to research was attractive since the
      &   travel agencies may be more creative and the library will
   1.3.1.1   have a much more complete travel section.

A **The IMP System Marketing Team** produced the following Graphical Format PDPC. (See pages 196-197.)

Exhibit PD-5
Develop Product Tests Set in Both the Laboratory and Everyday Situations

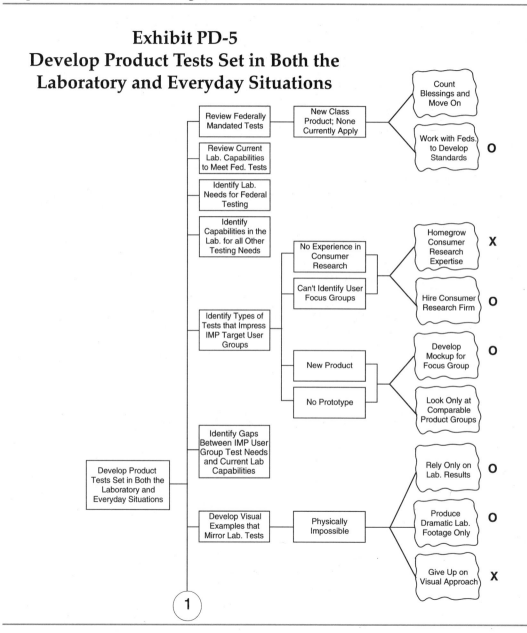

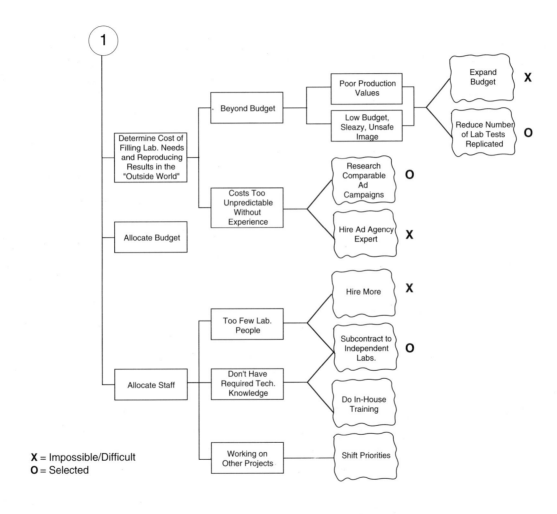

X = Impossible/Difficult
O = Selected

Based on the IMP System PDPC, the following decisions were made:

In Advance
- To hire a consumer research firm.
- Develop mock up for focus groups.

Response to Contingency
If necessary to:
- Work with Feds to develop standards if none exist.
- Rely on dramatic lab footage if replicating the test is possible.
- Reduce the number of lab tests replicated if too expensive.
- Subcontract to independent labs if Frankel lacks the number of people or technical skills.

Summary

In the end, the PDPC is consistent with the other Management and Planning Tools in that it makes breakthrough possible by looking for simple answers to simple questions. These answers to repeated "what if's" add up to a very complete implementation picture. This "completeness" is the key. It is the equivalent of turning over the last stone to see what crawls out. All of us have experienced that reluctant feeling of not wanting to raise the "killer questions" as the planning process winds down. It is that reluctance that has caused flawed plans to be launched.

The PDPC process, regardless of the specific format that it follows, encourages the "killer questions" to fall onto the table for discussion (like a skeleton falling out of a closet). If managers can have an accepted process for raising such questions as a matter of course, there may be fewer wounded messengers as well as plans that never should have been.

As with the remainder of the tools, the key is patience and trust in the process. It works!

Other Sources of Information

There are other resources that can help you learn about the uses and techniques of constructing the PDPC, or how to *facilitate* the use of the PDPC, as well as other management and planning tools. Several of these resources are:

- *The Memory Jogger™ II*
- *Coach's Guide to The Memory Jogger™ II*
- *The Coach's Guide* Package
- *The Learner's Reference Guide to The Memory Jogger™ II*
- *The Educators' Companion to The Memory Jogger Plus+®*
- *The Memory Jogger Plus+®* Software
- *The Memory Jogger Plus+®* Videotape Series

— *Notes* —

Chapter 7

Activity Network Diagram*

Definition

This tool is used to plan the most appropriate schedule for the completion of any complex task and all of its related sub-tasks. It projects likely completion time and monitors all sub-tasks for adherence to the necessary schedule. This is used when the task at hand is a familiar one with sub-tasks of a known duration.

The Activity Network Diagram is one tool that is certainly not Japanese. It is another name for a combination of the Program Evaluation and Review Technique (PERT) and the Critical Path Method (CPM) that were developed in the Operations Research discipline. It is simply one method of drawing the PERT/CPM relationships among activities in any project.

*This is simply a renaming of the "Arrow Diagram" used in the original Japanese text on the "7 New Tools." The author chose this name because it includes all of the formats available for showing a network of activity, not only the traditional Arrow Diagram method. This includes the Node Diagram/AON and Precedence Diagramming as well. These will be explained later in the chapter. The author also decided not to call it PERT/CPM because one of the purposes of *The Memory Jogger Plus+®* is to make these tools approachable by mainstream personnel, not just technical specialists. PERT/CPM has a rich but highly specialized history.

Activity Network

History of Activity Network Diagram/PERT/CPM

The roots of the Activity Network Diagram go back at least to the early 1930s in the work of a Polish scientist named Karol Adamiecki.[1] He developed an L-shaped matrix with a time scale on the vertical and the activities listed on the horizontal. This model, known as a Harmony graph,[2] featured moveable strips under each activity. The length of each strip indicated the duration of that activity on the vertical time scale. It had two noteworthy features:

1. The horizontal axis also included which activities were dependent on each other.

2. The strips were moveable to use the chart to accurately show the current status of the entire job.

This useful tool lay dormant until resurrected in 1958 when a research team from Lockheed Aircraft Corporation, the Navy Special Projects Office, and the firm Booz, Allen and Hamilton collaborated around a formidable task: How to bring the announced Polaris missile system, a major link in the nuclear deterrent plan, to deployment as quickly as possible and within budget. This was the most integrated weapon system to date and smaller scale projects were found to exceed completion and cost estimates by 40-70%.[3] Such "off target" results had deep strategic implications. PERT[4] provided estimates of pessimistic, optimistic, and likely times to completion that combined to provide a measure of probability for project success.[5]

---

[1] This and other historical details are drawn from the excellent text *Project Management with CPM, PERT and Precedence Diagramming* by Moder, Phillips and Davis, Van Nostrand Reinhold Co., New York, 3rd Edition, 1983, pp. 10-14.

[2] Ibid, p. 11

[3] Ibid, p. 12

[4] According to Moder, p. 12, "PERT" was originally the name of the project, "Program Evaluation Research Task." "Task" later became "Technique."

CPM (Critical Path Method) was being developed almost simultaneously to PERT (1956-59) by a joint du Pont/Remington Rand Univac team. Its target was to develop a method to trade off time to completion and cost in well defined projects like plant overhauls, maintenance, and construction. CPM accomplished this by identifying the shortest possible time within which the project could be completed based upon the longest path of activities. Project managers could then review each of the jobs along this path and determine the cost to do any of them more quickly.

PERT and CPM were not used together until after the completion of both projects. Since then, they have been combined to the point that they seem almost inseparable techniques. They have both been modified and refined to take full advantage of developments in computer technologies.

Various Construction Options[6]

These are three major ways to represent the output of the PERT/CPM process. All three are optional forms of the Activity Network Diagram.

1. Activity On Arrow: In this method, the arrows represent activities of some duration and the nodes (usually circles) represent the instantaneous start and finish of these activities. It is the format that was used in the original PERT research back in the 1950s. This is the "standard" Arrow Diagram.

---

[5] Similar developments were occuring in Great Britain, France, and elsewhere in the United States, Moder, pp. 12-14.
[6] Special thanks to Richard Zultner of Zultner & Co., 15 Wallingford Drive, Princeton, NJ, 08540-6428, for his help in understanding the advantages and disadvantages of these options. Much of the material in this construction option section is derived from his unpublished writing on this subject.

This is the method that will be utilized in the next section of this text to illustrate the various terms and calculations used in the PERT/CPM.

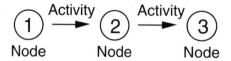

2. Activity On Node/AON or Node Diagram/AON: The method in which the nodes (usually rectangles) represent activities of some duration and the arrows simply the precedence of those activities. Since being developed in the late 1950s (shortly after the Arrow Diagram) it has become more and more widely used, especially in Europe. This was encouraged by the development of computer programs to accept and support this format in the mid 1970s.

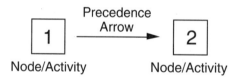

3. Precedence Diagram/PDM: This refinement of the Node Diagram allows the arrows between the nodes to represent leads or lags. In brief, both the Arrow and Node Diagrams assume that every job is <u>completely</u> finished before its successor can be started. The PDM allows the user to show, for example, how jobs can be started in a construction project when its predecessors are only 25% completed, e.g., smoothing concrete once it has been poured in a driveway or walk.[7] In fact, in this concrete example, to start the smoothing only after the pouring has been finished would create terrible quality problems. Both the Arrow Diagram and the Node Diagram have difficulty showing this type of partial completion time lag.

---

[7] *A Management Guide to PERT/CPM with PERT/PDM/DCPM and Other Networks,* by West and Levy, Prentice-Hall, Inc., Englewood Cliffs, NJ, 2nd ed., p. 136.

Common Language and Symbols Used in
Activity Network Diagram/PERT/CPM[8]

In order to understand the steps in the construction of these techniques, we must know the most common symbols and terminology. The example will use the Activity on Arrow (Arrow Diagram) format.

1. <u>Event, Node</u>: This is the beginning and the finish of a job and each is the connecting point to another job. Each has a unique number for ease of reference.

    Example: "Node 1"

2. <u>Job, activity</u>: This is the task that requires a length of time for completion. It is referred to by the node numbers that mark its beginning and end.

    Example: "Job 1, 2"

3. <u>Immediate Predecessor</u>: A job that must be completed before another job can begin.

    Example: "Job 1, 2 as Immediate Predecessor of both Job 2, 3 and Job 2, 4"

4. <u>Immediate Successor</u>: A job that cannot be started until another job is finished.

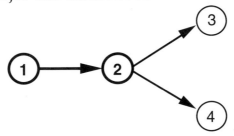

---

[8] The terminology and symbols used are fairly standard but the major reference for these is *A Management Guide to PERT/CPM with PERT/PDM/DCPM and Other Networks*, by West and Levy, Prentice-Hall, Inc., Englewood Cliffs, NJ, 2nd ed., pp. 5-40.

Example: "Job 3, 5 and Job 4, 6 as Immediate Successors of Job 1, 2"

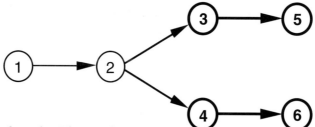

5. <u>Initial node, terminal node</u>: This indicates which nodes start a job and which nodes end a job.

Example: "Node 1 is Initial Node for Job 1, 2; Node 2 is the terminal node of Job 1, 2 and the Initial Node of Job 2, 3 and Job 2, 4"

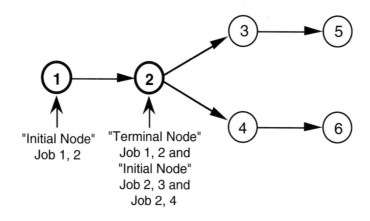

6. Duration: The length of time required for the completion of any job.

Example: "The duration of Job 1, 2 is 3 days, Job 2, 3 is 2 days, etc."

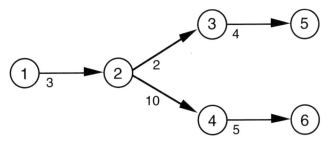

7. Dummies: Symbols used to simplify the PERT Chart when one job has more than one node feeding into it from the same immediate predecessor. It prevents confusing double arrows and jobs with the same name (e.g., two Job 2, 3's).

Example: "Before Job 3, 5 can happen both Job 2, 3 and Job 2, 4 must be completed. PERT by itself cannot show two jobs coming from the same node, feeding into the same node."

Without a Dummy:

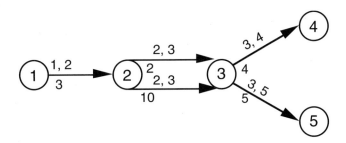

With a Dummy:

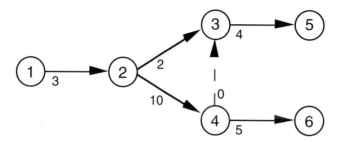

8. **Earliest Start Time (ES):** Given the cumulative duration of all of its predecessors, the earliest any job can start.

Calculation: Maximum of the duration of all predecessor jobs. This is the longest implementation path feeding into the job in question. If the Earliest Finish (EF) has been calculated this can be expressed as Max (Earliest Finish of any of its predecessors).

Example: "Job 2, 3 and Job 2, 4 cannot start any earlier than 3 days into the project; Job 3,4 and Job 4, 6 cannot start any earlier than 13 days into the project; Dummy Job 4, 3 cannot start until 13 days into the project."

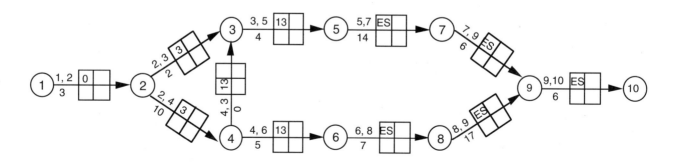

9. <u>Earliest Finish Time (EF)</u>: Given the Earliest Start Time (ES) and the duration of the job under question, the earliest any job can be completed.

Calculation: Earliest Start Time of each Job + the Duration of that Job.

Example: "Job 1, 2 cannot finish any earlier than 3 days into the project; Job 2, 3 cannot finish any earlier than 5 days into the project; Job 2, 4 cannot finish any earlier than 13 days into the project; Dummy job 4, 3 cannot finish any earlier than 13 days into the project; Job 3, 5 cannot finish any earlier than 17 days into the project; Job 4, 6 cannot finish any earlier than 18 days into the project."

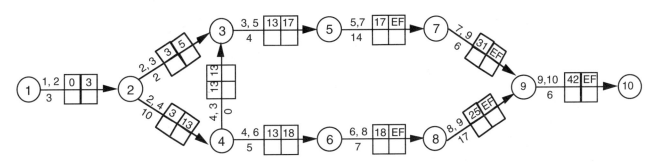

10. <u>Latest Finish Time (LF)</u>: Given the job(s) that any job must feed into, the latest that the job in question can finish without delaying the start of the job where two or more paths merge. Finishing any later would result in delays in the overall project schedule.

Calculation: Starting with the last job in the Project apply Total Duration (Critical Path/Earliest Total Finish)—the duration of the last job. This gives you the Latest Start of the last job. All predecessors use this as the basic calculation. Each job's latest finish time therefore equals the latest start of each job's immediate successor.

Example: "Job 9, 10 must start no later than the 42nd day of the project. Therefore, in order to stay to the schedule both Jobs 7,9 and 8,9 must finish no later than the 42nd day."

11. <u>Latest Start Time (LS)</u>: Given the job(s) that any job must feed into, the latest that the job in question can begin without delaying the start of the job where two or more paths merge. Missing this start time would result in delaying the immediate successor's start.

Calculation: Latest Finish Time of each job — the duration of that job.

Example: "Job 2, 3 must start 17 days into the project or the total schedule will be delayed; Job 2, 4 must start 3 days into the project or the total schedule will be delayed; Job 3, 5 must start 19 days into the project or the total schedule will be delayed; Job 4, 6 must start 13 days into the project or the total schedule will be delayed."

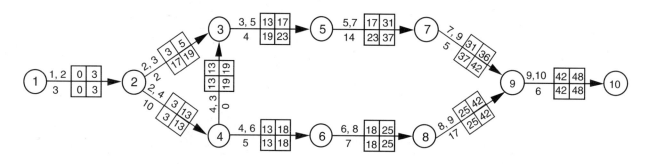

12. Critical Path (CP): The implementation path with the longest cumulative duration. This is the quickest that the total project can be completed. Any delay in jobs along this path will automatically result in the delay of the total project.

   Example: "The path including Jobs 1, 2; 2, 4; 4, 6; 6, 8; 8, 9, and 9, 10 is the Critical Path (CP)."

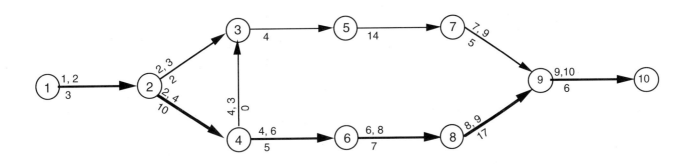

13. Total Slack (TS): The amount of time that any job can be delayed without delaying the overall project schedule. By definition, all those jobs on the Critical Path have zero slack since any job delay will cause project delays.

   Calculation: Once ES, EF, LS, and LF have been calculated for each Job, TS = Latest Start — Earliest Start of each Job, or, TS = Latest Finish — Earliest Finish of each Job.

   Example: "Job 1, 2, Job 2, 4 and Job 4, 6, etc. are on the Critical Path and have zero slack. Job 2, 3 has a slack of 14 days, Job 3, 5 has a slack of 6 days."

(See Item 11 above.)

14. <u>Forward Pass</u>: The process of calculating all of the earliest starts and finishes, starting with the first jobs and working <u>forward</u> to the final jobs in the project sequence.

  Example: See items 8 and 9 above.

15. <u>Backward Pass</u>: The process of calculating all of the latest starts and finishes, starting with the last jobs and working <u>backward</u> to the first jobs in the project sequence.

  Example: See items 10 and 11 above.

Activity Network Diagram Construction Options[9]

	Advantages	Disadvantages
Arrow Diagram	• Familiar • Arrows suggest actions, motion • Well suited for computers because the two number job identification automatically supplies all the connections of any job to all of its successors • "Events" orientation consistent with language of PERT	• Requires dummies to be used
Node Diagram/AON	• Simple to draw • Simple to explain • Easy to understand by non-technical users • Eliminates the need for dummies	• Not as familiar
Precedence Diagram Method/PDM	• Shows overlapping activities accurately by allowing for more options beyond the traditional "must finish to start" format • Eliminates the need for dummies	• Not as familiar • Appears more complex because less cut and dry

[9] Many of these evaluations are included in West and Levy, pp. 13-15.

When to Use an Activity Network Diagram

The most important criterion (and perhaps the only meaningful one) is that the subtasks, their sequencing, and duration must be well known. If this is not the case, then the construction of the Activity Network Diagram can become a very frustrating experience. When the timing of the actual events is so different from the original diagram, people dismiss it as an exercise that stresses form over substance. After such a negative experience, rest assured that it will not be used again voluntarily. When there is a lack of history about an implementation process, the PDPC may provide more insight into a likely implementation schedule. What the PDPC lacks in precise scheduling detail is balanced by its realistic picture of the task that must be implemented.

Given these factors, the Activity Network Diagram is most useful when:

- The task is a complex one.
- The subtasks are familiar ones with known durations even if they may have been combined in different sequences in the past.
- The task at hand is a critical organizational target.
- There are simultaneous implementation paths that must be coordinated.
- There is little margin for error in the actual vs. the estimated time to completion.

Typical Uses of an Activity Network Diagram

- An electronics manufacturer is developing a new PC board construction process. It involves the coordination of hardware, software, and raw materials development as well as the training of the personnel who will run and maintain the time.

- A theater management company must coordinate the timetable of constructing a new theater with the construction schedule of the new shopping complex in which it is located.

- A Director of Training must propose a training and education schedule to support a TQM Master Plan. It entails the integration of instructors, facilities, materials, and promotion with the program rollout that it must support.

- A training organization must shorten the amount of time that it takes to publish its yearly catalogue of courses. This requires input on instructors' availability, new course development, and facility scheduling.

Construction of an Activity Network Diagram[1]

 Assemble the Right Team

Since the Activity Network Diagram must produce a credible implementation schedule, it requires the most intimate knowledge of any of the tools. Therefore, the team should include managers and employees as close to the situation as possible. At the very least, the managers on the team should get the input of employees as appropriate. The last thing you want to do is simply assign it to a specialist for development in isolation. Remember that this tool has been around for over thirty years in the hands of "experts." The purpose of including the Activity Network in this book is to help make it more "user friendly" to mainstream managers.

B[2] The Missed Promised Delivery Dates Team

Based upon the team's analysis through the Matrix Diagram, their focus shifted to "Shortening the 11-digit product code." The team consisted of:

---

[1] For illustration purposes, the Activity on Node/Node Diagram will be used in the Missed Promised Delivery Dates Team example.

[2] In the Frankel case study, the A Team (IMP System Marketing Team) chose to do a PDPC instead of an Activity Network Diagram (see Chapter 6). Please note that the tasks in a PDPC, including countermeasures, could be carried over to an Activity Network Diagram to plan and monitor an implementation schedule.

- The MIS manager
- The data entry supervisor
- A production supervisor

This team put together the basic flow of the implementation steps but received invaluable input from several hourly employees in the shipping and data entry departments.

2 Brainstorm and Record All of the Tasks Needed to Complete the Project

The team must determine all of the project's necessary tasks by using either a brainstorming or data gathering process. Brainstorming is appropriate when the team has thorough knowledge of the project or a breakthrough is required in implementation. "Data" could also be gathered from past similar implementations to get the complete list of tasks.

Whether original brainstorming or a review of past implementations is done, follow these process steps:

a) Record the jobs simply and clearly on cards (3M Post-it™ Notes are recommended). This is essential for moving the jobs around before the final plan develops and lines and arrows are drawn in.

b) Divide each card (referred to as a "job card") into two sections.

2 **Brainstorm and Record All of the Tasks Needed to Complete the Project** (Cont'd.)

Record the job on one half of the card. The length of time to complete that job will be filled in this space later. For example,

Job 1

Job 2

B **The Missed Promised Delivery Dates Team**

In a two-hour brainstorming session, the team generated the following fifteen tasks.

Identify software coding for product code processing

Identify overlapping current customer uses of product code

Identify current product code "customers"

Identify software problems for current product code users

| Identify current customer uses of product code | | Identify software capability upgrade required for ideal customer uses | |

| Identify current customer product code problems | | Identify overlapping uses of product code info. from the ideal customer needs | |

| Develop ideal customer product code needs beyond current uses | | Develop software development budget | |

| Prioritize all product code customer needs | | Implement product code software in test locations | |

| Review and finalize software | | Develop product code software | |

| Select final list of customer product code uses | |

Sequence All of the Identified Activities

Once all of the necessary tasks are identified, then two types of flows must be determined:

a) <u>The Sequential Flow</u>: This is the relationship among all the jobs regarding which tasks precede and follow each other. This determines the immediate predecessors and successors of each job.

b) <u>The Simultaneous Flow</u>: This identifies those tasks that are done in parallel to other tasks.

Begin the process by identifying the jobs that form the longest path of sequential tasks.

NOTE 1: Use cards or Post-it™ Notes to ensure that the sequencing can change as needed.

NOTE 2: Feel free to delete duplicate jobs and add new ones as necessary.

Next, identify the second longest sequence of jobs that must be done to support the main implementation path and that can be done simultaneously. This process (using the cards or Post-it™ Notes) is repeated as many times as is required to include all of the jobs in some straight line path. For example:

3 Sequence All of the Identified Activities (Cont'd.)

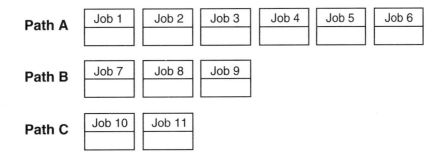

Next, identify the places in all of the paths where there are necessary connections with other paths. In other words, are there jobs in one implementation path that can't be started until key jobs in the parallel path are completed? For example:

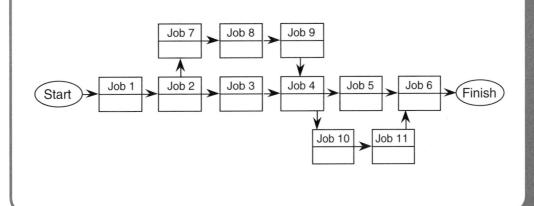

B The Missed Promised Delivery Dates Team

The team produced the following draft diagram.

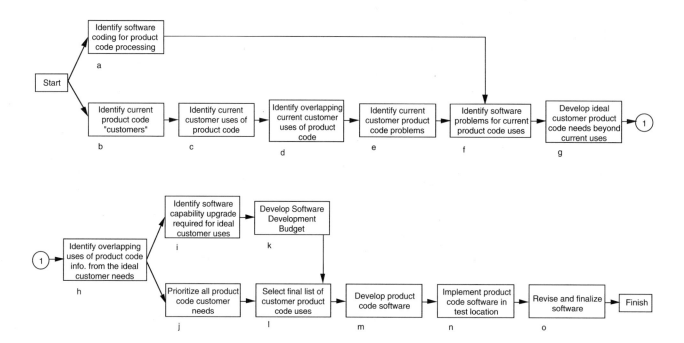

Give a Time Duration to Each Job

The team must now carefully examine past performance of all the tasks in order to estimate their probable duration in this project. The number of hours, days, or weeks it takes to complete a specific job often doesn't change from project to project. Record the duration on each job card.

NOTE: This type of information assumes that there have been accurate project records maintained over time. Absent this, the team is dependent on the experience and memory of former participants. This is clearly not foolproof.

B The Missed Promised Delivery Dates Team

After consulting with the representatives of both Software Service and Production Planning, the team arrived at durations for each job.

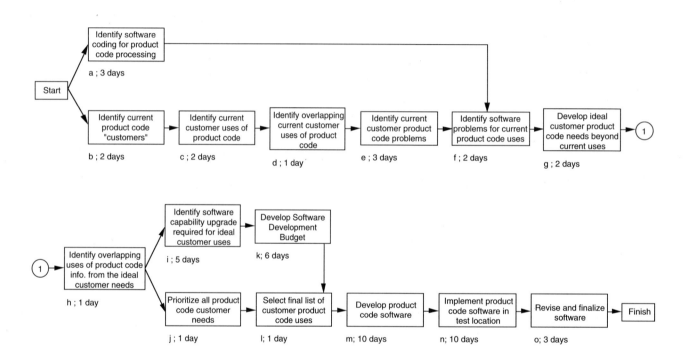

Calculate the Shortest Possible Implementation Schedule Using the Critical Path Method

The team must now calculate the shortest time within which the project can be completed based upon the longest cumulative duration of any path. This is the project's "critical path."

Practically speaking, it is important to know this for three reasons:

1) It reveals whether or not the timetable goal (if there is one) is achievable. For example, we state that the project must be done in 2 weeks but the critical path adds up to 2-1/2 weeks.

2) It identifies those jobs along the critical path that have no slack. They must be done on time.

3) It identifies the jobs along the critical path as the ones that must be done more quickly if the goal is to be achieved. These become the target of additional resources and/or creative solutions.

B The Missed Promised Delivery Dates Team

The team identified the following path (in **bold**) as their project's critical path.

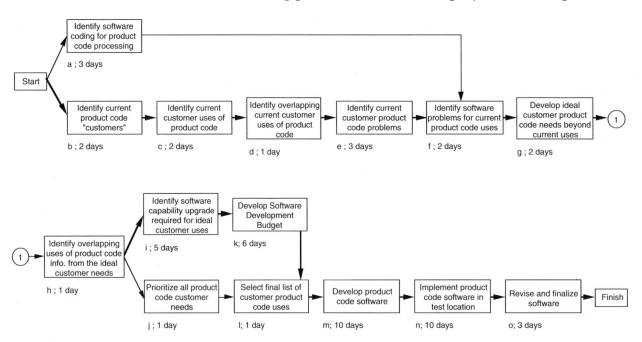

This showed that the new, shorter product code number would take 48 working days to implement. Furthermore, the schedule was a very tight one in that out of fifteen jobs only two were not on the critical path. This means that virtually all the jobs have to be completed as scheduled or the job will not meet the projected completion date.

Upon reviewing the critical path, the team agreed that the total implementation time was within (just barely) the 10 week implementation target that had been originally set. Therefore, it was decided not to devote more resources to the project in order to shorten the duration of the jobs along the critical path.

Calculate the Earliest Starting and Finishing Times and the Latest Starting and Finishing Times for Each Job

The team must now identify the starting and completion times for each job that is required in order to fit the jobs into the critical path schedule. This is done using a combination of the Forward Pass (see below for more details) and the Backward Pass (see page 228 for more details).

The Forward Pass: This process looks at each job starting from the beginning of the Activity Network Diagram. For every job, use the cumulative duration of all predecessors of that job plus the duration of that job to calculate its Earliest Start and Earliest Finish.

For example,

This is repeated for each job in the network.

Calculate the Earliest Starting and Finishing Times and the Latest Starting and Finishing Times for Each Job (Cont'd.)

The Backward Pass: This process looks at each job starting from the end of the Activity Network Diagram. Begin with the total implementation time of the entire network. This becomes the latest finishing time of the last job of the network. Subtract from that total the duration of the final job.

For example,

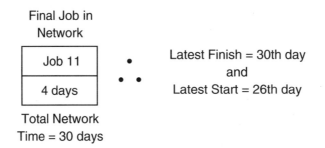

Continue the backward pass by making the latest start of the final job the latest finish of all its immediate predecessors. Subtract the duration of each job from its latest finish to get its latest start. Repeat this sequence for every job.

6 Calculate the Earliest Starting and Finishing Times and the Latest Starting and Finishing Times for Each Job (Cont'd.)

For example,

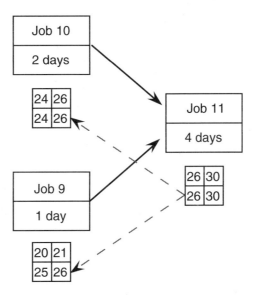

NOTE: The most common way to summarize the Earliest Start/Finish and Latest Start/Finish is

ES	EF
LS	LF

B The Missed Promised Delivery Dates Team

Once the critical path was identified, the team then did both the forward and backward pass process with the following results.

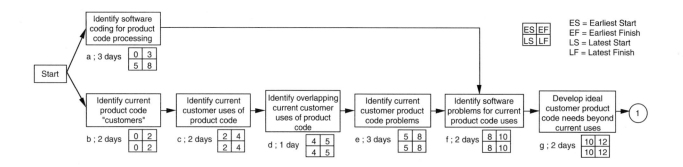

ES EF
LS LF

ES = Earliest Start
EF = Earliest Finish
LS = Latest Start
LF = Latest Finish

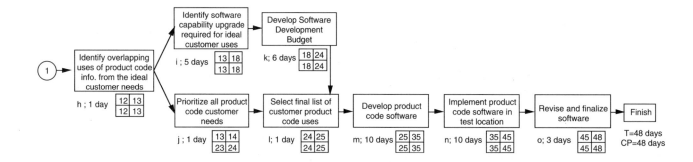

Locate Jobs With Slack Time and Calculate Total Slack

To find slack time in any job, simply review each node for differences between the earliest start (ES) and latest start (LS). Where there are differences, subtract the earliest start from the latest start to calculate that job's slack time. Total slack (TS) for the entire network is simply the total of all individual slack times across all the jobs in the network.

NOTE : By definition, all jobs on the critical path will have no slack. Therefore, these jobs must meet the schedule without deviation if the overall schedule is to be met.

B The Missed Promised Delivery Dates Team

The team concluded that there were actually 15 days of total slack in the network. Some felt that this provided a healthy "cushion." They quickly discovered, however, that all of that slack existed in only two jobs. Every other job was on the critical path! They decided to closely monitor the critical path and shift resources to the critical path job whenever they came across a job with some slack. Further, they decided to not use more than half of any job's slack time so that those jobs would not become critical jobs as well.

8 Review and Revise the Activity Network Diagram

As with all the 7 MP Tools, it is wise to review the final product with those whose work will be monitored by it. Above all, be prepared and open to change. If feedback is not heard, considered, and incorporated where needed, people quickly become cynical and the diagram as a planning tool may become less and less useful.

Summary

In the end, the Activity Network Diagram (both Arrow and Node formats) is yet another tool to make the Plan-Do-Check-Act (PDCA) cycle come alive. It is not intended to drive people, but to signal them when the plan is not going according to schedule. This is an opportunity to diagnose the problem, make some changes, and continue to monitor the schedule impact of those changes. It is also an opportunity to assess the present scheduling performance with an eye toward breakthrough. If the projected total network implementation is not satisfactory, then what in the system will change to shorten it?

Finally, the Activity Network Diagram in all of its forms has often been relegated to only the most complex projects. The key is to look for applications in the "simpler," smaller scale projects that often suffer from poor schedule performance. We must again put this technique along with the six other 7 MP Tools into the hands of the "new planners": mainstream managers at all levels.

Other Sources of Information

There are other resources that can help you learn about the uses and techniques of constructing the Activity Network, or how to *facilitate* the use of the Activity Network, as well as other management and planning tools. Several of these resources are:

- *The Memory Jogger*™ *II*
- *Coach's Guide to The Memory Jogger*™ *II*
- *The Coach's Guide* Package
- *The Learner's Reference Guide to The Memory Jogger*™ *II*
- *The Educators' Companion to The Memory Jogger Plus+*®
- *The Memory Jogger Plus+*® Software
- *The Memory Jogger Plus+*® Videotape Series

—*Notes*—

Chapter 8

Implementing the Seven Management and Planning Tools: Behavioral Requirements

Since working with the "Seven Management and Planning Tools" (7 MP Tools) in 1984, we have found them to have a powerful effect on how teams approach very challenging problems:

- Make consensus achievable in a non-threatening style
- Help uncover breakthrough
- Reveal gaps in logic
- Let creativity combine with very thorough implementation planning

Like any other tool, however, they are only as good as the user applying them. A saw in the hands of a finish carpenter is like an artist's brush. Give that same saw to a seven year old child and just watch your dining room table get several inches shorter.

The 7 MP Tools not only require technical skills (the steps, appropriate uses, do's and don'ts, etc.) but behavioral skills as well. The following can be used as a personal checklist. Use it to evaluate your <u>preferred</u> style of planning and problem solving. If you find yourself coming up short on a good percentage of the points, you will likely feel uncomfortable using some or all of the tools. Judge for yourself.

Behavioral Requirements

1. Trust The Process

Because these tools are often unfamiliar and sometimes appear complex, participants are often in unfamiliar territory. We often want to know exactly where we are in the process, how much more is to be done, and how it all relates to the goal of the process.

The Japanese seem to treat this as a non-issue. They have a great confidence in ritual. They are willing to suspend judgment until the steps of the ritual are completed. Then they step back and see what emerged. We refer to this as "emergent thinking."

Often in the West, we seem suspicious of anyone who says "Trust me, the results will be worth all of the effort." It may be this same tendency that makes it difficult for experienced production people to "let the process speak" through statistical process control, etc.

This suspension of judgment in process is the key to really achieving breakthrough in thought. Otherwise, we often steer the process to make our conclusions self-fulfilling prophecies. It's only through this trust of the process that new ideas can rise to the surface through our filters of bias, opinion, and experience.

Like a ritual, the 7 MP Tools may appear to be complex, but closer study reveals that they consist of very simple steps followed in a disciplined way.

2. Value Brainstorming

"Brainstorming" has become a cliché. It has come to mean any generic process of creating a set of ideas for further discussion. To really value brainstorming one must really believe:

- That the whole is greater than the sum of its parts
- That a team can generate a large number of very creative ideas very quickly
- That each idea can serve as a catalyst for many more
- That no idea can be thrown out

This also assumes that ideas are as legitimate as data. The Japanese often refer to the raw materials of the 7 MP Tools as "idea data." We must learn to combine the power of ideas and data and treat each as a resource.

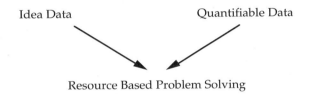

3. Discipline

"Simple Things Done Perfectly"

The 7 MP Tools, like many Japanese systems, appear to be complex and extremely detailed. Also, as in most Japanese systems, the appearance of complexity is only a composite of many simple steps done right each time. For example, the Interrelationship Digraph sometimes looks like a road to chaos, but it results from one question being asked repeatedly: "Does this item cause/influence any other item?" It's not unusual to have 20 x 20 matrices which represent 400 relationships to be considered. But each matrix, no matter how large, is still the result of the same disciplined questioning: "Is there a relationship between these 2 or 3 items?"

The challenge is to see those building block steps as worth doing and worth doing well: Discipline.

4. Patience

"Patience may be a virtue,
but it sure won't get you to the top."

Many of these behaviors are interrelated. One of the "root behaviors" may be the lack of patience in most American management systems. Most managers don't feel that they can take the time to wait for a long process to play itself out. Thus, there is a built-in bias against digging for root causes. This often extends to planning itself. Without personal and organizational patience, the 7 MP Tools may appear too countercultural.

5. Trust in Initial Gut Reaction

For tools such as the Affinity Diagram it's extremely important not to over-analyze. Creative solutions come from new combinations of ideas. Until we learn to have confidence in these new combinations, we will always preserve the status quo. Some users fear that these new combinations will be too "crazy." Rest assured that these tools have enough checkpoints to weed out the truly "off the wall" schemes.

6. Listening Skills

Simply stated, the 7 MP Tools stress consensus; consensus requires understanding basic issues, causes, and solutions; this understanding requires each team member to keep open ears and an open mind. Without good listeners on the team, the entire process breaks down and creates mediocre status quo solutions and plans.

7. Common Sense to Know When to Stop

As helpful as these tools have proven to be, there is a danger of overuse. A classic example is the Tree Diagram. It can consist of two or ten levels of detail. At some point, a task is assignable and doesn't require further breakdown. Moreover, to explode a tree into too much detail can remove the incentive to innovate at the lowest implementation level.

8. Common Sense to Know When the Tools are Not Appropriate

Number 7 above referred to overuse by pushing one or more of the 7 MP Tools further than necessary. This point is overuse by force-fitting a tool where it doesn't work. This happens primarily when a 7 MP Tool enthusiast takes a simple implementation of a common task and burdens the plan with unneeded analysis. In this case, not only is the original plan needlessly delayed, but the tool that's force-fit gets a bad name because it doesn't add value.

If you only remembered one guideline from this book let it be this:

When you're facing a formidable task, either because it is:

- *So new it is ill defined*
- *So old it's deemed impossible*
- *So critical that it scares everyone away*

...Give the 7 MP Tools a try!

9. Integrity: People Commit to Use the Outcome of the Process

Make no mistake that the 7 MP Tools involve more <u>structured</u> planning time than is customary. Even if the tools produce tremendous results, members of the team know that they've spent a fair amount of time in the process. Now take that same team, go through the process, dismiss the results, and then declare it a "learning experience" and you have a disaster. Probably their biggest learning is that "It just ain't worth it."

Granted, there will be a learning process with each use of the tools, but the emphasis must always be: We have an important problem, we have the right people together, and now let's use the 7 MP Tools to get the best possible solution. Even an imperfect use of the tools will be an improvement over many current planning practices.

10. Flexibility, Tolerance of Ambiguity, and Creativity

All three behaviors are essential because of one fact:

The 7 MP Tools have very few rules of the road.

The process often takes unpredictable twists and gets more than an occasional roadblock. A person with a strong need for predictability at all times may feel uncomfortable and directionless. Likewise, a person who simply gives up when an obstacle to progress appears, lacks the creativity to find the innovative path.

11. Value, Not Simply Tolerate, the Different Perceptions of Others

Participative management is in vogue today. In most progressive organizations there is also a strong emphasis on teamwork. Despite this encouraging movement, many managers are still "going through the motions." They know they <u>should</u> seek input, encourage differences of opinion, and get as many views of a situation as possible, but being the Lone Ranger is easier (and more fun). If the 7 MP Tools are to become commonly used, then a manager must see them as a helpful mechanism to seek and get the input of all team members, from diagnosis to implementation.

— *Notes* —

Chapter 9

Implementing the Seven Management and Planning Tools in Your Organization

Training

Many people that we have trained have been tempted to return to their organization and plug the 7 MP Tools into their Management Training Curriculum. Simply stated, put every manager "through a course." This appears to be a worthwhile idea, but it has some serious disadvantages:

- Viewed just as more tools in an already overflowing toolbox
- Not clear how it will be of practical help to my daily work
- Viewed as too Japanese
- The entire cycle is too complex, no time to do it all

The reason that all of these are serious problems is that the 7 MP Tools are learned best experientially. They require <u>PRACTICE</u> above all else. They require more practice than most courses allow. Furthermore, simulated applications, unlike an SPC course where data can be supplied, often create unpredictable results. If people are to be convinced of the 7 MP Tools' usefulness, the application must be very relevant to their jobs, or it's tough to know when a breakthrough has occurred. Without this "AHA!!" feeling, someone could walk away instead saying "So what?"

What's the alternative then to the traditional classroom approach? The following are a few suggestions:

Implementation

A. Opportunistic Training

1.) Send several managers to an outside course to pick up the basics of the 7 MP Tools. These managers should be:

- <u>Interested in new planning methods</u>. Go where the interest is. Don't send them to be convinced.
- <u>At a high level of responsibility</u>. The ideal candidate is someone at a department manager level or higher. In other words, someone with significant planning responsibilities.
- <u>Credible with the top management of the organization</u>. These managers should be able to "suggest" a new approach to top managers.
- <u>Line managers, not staff specialists</u>. Although most of the progress to date has been due to the hard work of staff specialists, they have to overcome some attitudes to do so. For example, many managers may feel that staff people don't know the "real world"; that they are coming around again with more "new stuff."

2.) Have that line manager go through 1-2 projects using the 7 MP Tools in his or her own job. This allows the manager to:

- Build confidence
- Modify the tools to fit the organization's needs
- Build some internal case studies
- Build a little lower level support and knowledge
- Identify the "critical few" tools that seem to work best in the organization's culture

It will probably take 2-3 months to accomplish this "R & D" effort.

3.) Train either the Top Management Staff or the Total Quality Steering Group.

In either case these are the "do or die" students. They are responsible for the most comprehensive planning tasks in the organization. The kinds of issues they must resolve are often complex, deep, and critical. Furthermore, if they don't see the usefulness of the 7 MP Tools, then no one will place the demand on the system to use them.

4.) Pursue more "Opportunistic Training."

- When individually trained managers see applications, encourage them to introduce one or more of the tools. This should be a "Why don't we try this approach (e.g., Affinity Diagram)?" vs. "Let me teach you about this break-through Japanese planning tool called the Affinity Diagram!"
- Provide training to new project members as they get started, particularly if they have a Total Quality focus.
- Include a process facilitator in new project groups who knows the 7 MP Tools well. When the group is at a critical crossroads, the facilitator can suggest and facilitate the use of one or more of the 7 MP Tools.

The purpose behind this opportunistic training approach is to present the 7 MP Tools when they are appropriate. If they actually help a team to progress through a difficult issue, the tools become a part of each member's skill set. They are more likely to be used quickly than if someone has to translate classroom knowledge to the "real world." Moreover, many of the managers who tend to use the 7 MP Tools extensively resist training. Training in specific management skills is often a delegated task to a subordinate. The ideal for this middle to upper management audience is for them not to even suspect that they are indeed being trained.

B. Classroom Training

There is one disadvantage to opportunistic training: it appears to be too slow. It can be argued that this is deceiving since the <u>retention</u> and <u>use</u> rates tend to be higher with classroom training. However, if there is a real need and strong management support for the 7 MP Tools, then organization-wide 7 MP Tools training can work. Such training would fall into two groups:

1.) <u>Full Cycle Training</u>: Approach in which all of the 7 MP Tools are taught as a single flow of activity. In this training model, all tools are taught with the output of one tool becoming the input of another.

2). <u>Individual Tool Training</u>: Approach in which any of the tools are singled out as helpful and therefore included in a broader curriculum.

1.) Full Cycle Training

Recommended model:

- 1-1/2 to 2 day course
- 20-30 people
- Class should be divided into groups of 4-6 people who come from the same function/department
- Groups should be preselected and instructed to bring a critical, tough issue that the team is currently facing
- Allow 1-1/2 to 2 hours per tool, including about 30-45 minutes maximum lecture and a generic practice exercise
- Training facility should be spacious in order to allow teams to practice comfortably
- Provide each team with an ample supply of 3M Post-it™ Notes

2.) Individual Tool Training

Recommendations:

- Any of the 7 MP Tools can be taught individually, but the Affinity Diagram seems to be the one that most companies are integrating into broader training. It is perceived as simple, easy to learn and apply often, and powerful for breakthrough at many levels.

In either case (Full Cycle or Individual), allow plenty of time for practice, use real issues for practice, and include, whenever possible, examples of 7 MP Tools used in your organization.

Training at Which Level?

Even though the 7 MP Tools seem to be most comfortable for middle to upper management, we have seen impressive applications in other levels such as technical or staff specialists and even at the hourly or line employee level. In fact, in some Japanese companies, we have seen some of the 7 MP Tools integrated into their Quality Circle training process.

In summary, regardless of who you train in any of the 7 MP Tools, stress the following:

- Make it experiential
- Make it relevant to team members' work
- Keep it simple and approachable
- Keep the model flexible

— *Notes* —

Appendices:

Appendix A: Decision Flowchart for Using the 7 MP Tools
Appendix B: The 7 MP Tools Integrated with the Basic QC Tools
Appendix C: *The Memory Jogger*™

Appendices

Appendix A:

Decision Flowchart for Using the 7 MP Tools

In the introduction of *The Memory Jogger Plus+®* (see page 7), there is a simplified typical flow of the 7 MP Tools. This reflects how the tools can be used in combination. Each tool can be (and often is) used alone with excellent results. How do you decide which tool to use at any one time? When can you stop the process to change course? The following flowchart provides guiding questions that allow the user to follow the typical flow (page 7) or choose the appropriate single tool at the right time. **Please note that the "Basic QC Tools" (Run Chart, Pareto Chart, etc.) are presented as an appropriate alternative and/or supplement. They are suggested as different, not inferior techniques.**

Seven Management Tools Process Flow

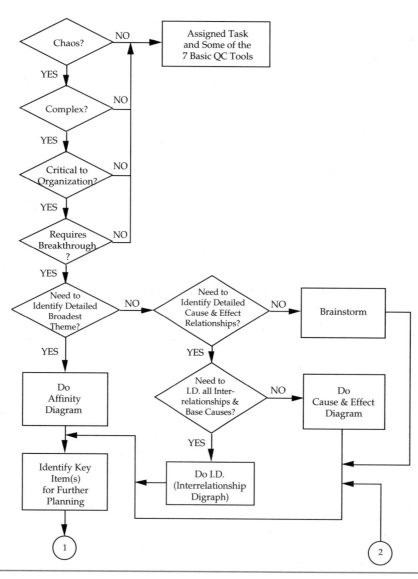

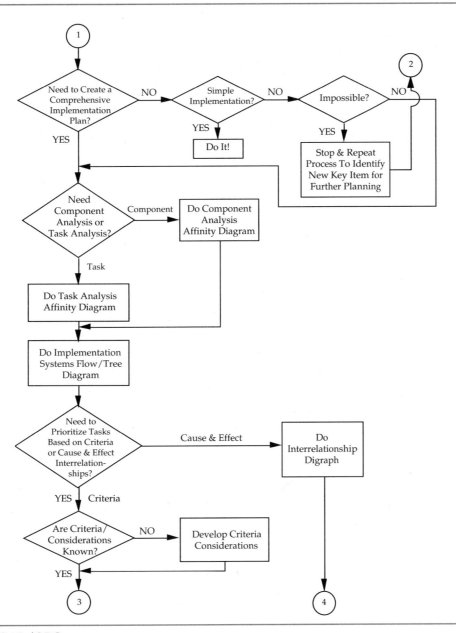

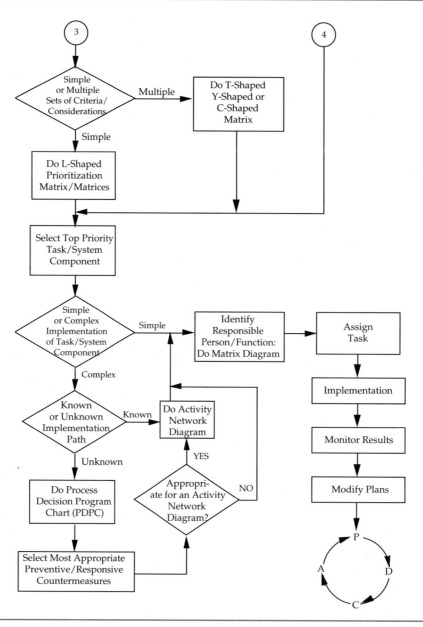

Appendix B:

The 7 MP Tools Integrated with the Basic QC Tools

The Japanese have shown us that both the 7 MP and 7 Basic QC Tools are at their most powerful when they are part of a structured problem-solving cycle. If they are to be most effective, these two sets of techniques should be seen as supplements, not replacements, of each other.

Another insight that the Japanese have provided is that any truly effective problem-solving process must balance the <u>data</u> and <u>knowledge</u> that surrounds the problem. It is not a matter of choosing between data and knowledge but combining them effectively.

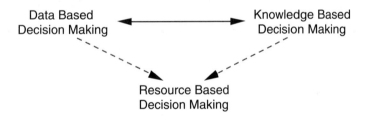

The graphic above illustrates the need to treat both data AND knowledge as resources to be used together. We must find a middle ground between two extremes of the problem solving continuum.

| "Don't talk to me unless you have the hard numbers." | vs. | "This is what I think. Don't confuse me with facts." |

The following problem-solving models combine both sets of tools in order to strike this balance.

A. Generic Eight Step, Problem-Solving Model: Track 1

This model includes a combination of Basic QC and 7 MP Tools in each step in order to accommodate situations that are either "hard," "soft," or a combination of the two.

Track 1: Generic Problem-Solving Model

Task 1: **Creation of Problem Area Master List**
- Flow Diagram/Imagineering
- Wide Open Brainstorming
  - Problem Based
  - Customer Demand Based
- Affinity (KJ®) Diagram

Task 2: **Problem Area Selection**
- Problem Data Collection and Analysis
  - Reports
  - Check Sheets
  - Run Charts
  - Pareto Charts
- QFD/House of Quality
- Interrelationship Digraph

Task 3:   **Problem Specification**
- Tree Diagram
- Confirming Data Collection

Task 4:   **Problem Definition/Description and Exploration of Possible Causes***
- Guiding Definition Questions
    - What the Problem **Is** and **Is Not**
    - Where the Problem **Is** and **Is Not** Happening
    - When the Problem **Is** and **Is Not** Happening
    - The **Extent** of the Problem
- Cause and Effect Diagram

Task 5:   **Determination of Basic Causes**
- Consensus
- Confirming Data Collection
    - Scatter Diagram
- Interrelationship Digraph

Task 6:   **Generation of Solutions and Action Plans**
- Tree Diagram
- Matrix Diagram
- Process Decision Program Chart (PDPC)
- Activity Network (Arrow) Diagram

---

*This is derived from the work of Charles Kepner and Benjamin Tregoe as outlined in their text *The Rational Manager: A Systematic Approach to Problem Solving and Decision Making*, McGraw-Hill, NY, 1965.

Task 7:   **Evaluation of Solutions and Action Plans**
- PDPC
- Activity Network Diagram
- Customer Survey

Task 8:   **Problem Solving Process Evaluation**
- Possibly Affinity Diagram

B. Reduction of Variability: Track 2

This problem-solving model, which parallels Track 1 mentioned previously, deals specifically with identifying and eliminating both special and common causes of variability. This process is by necessity, data-based. However, note in Task 6 that in order to reduce inherent variability, the entire range of tools may be used as in Track 1. This is based on the fact that the reduction of inherent variability (the area between the Upper and Lower Control Limits) is fundamentally systems improvement.

Track 2: Reduction of Variability Model

Task 1:   **Determination of Inherent Variability**
- Product/Process Sampling
- Control Charts
    - Variables
    - Attributes

Task 2:	**Determination of Special Causes**
	• Control Charts
	–Variables
	–Attributes
Task 3:	**Elimination of Special Causes**
	• Generalized Problem-Solving Track (Tasks 4-7)
Task 4:	**Assessment of Impact of Changes**
	• Control Charts
Task 5:	**Comparison of Process Performance to Specification Requirements**
	• Histogram
	• Capability Index
Task 6:	**Reduction in Inherent Variability**
	• Generalized Problem-Solving Track (Tasks 1-8)

C. Problem-Solving/Process-Improvement Model

There are many standard models for making improvements. They all attempt to provide a repeatable set of steps that a team or individual can learn and follow. The *Improvement Storyboard* is only *one* of many models that include typical steps using typical tools. Follow this model or any other model that creates a common language for continuous improvement within your organization.

Plan

Depending on your formal process structure, Step 1 may be done by a steering committee, management team, or improvement team. If you are an improvement team leader or member, be prepared to start with Step 1 *or* Step 2.

1. **Select the problem/process that will be addressed first (or next) and describe the improvement opportunity.**
   - Look for changes in important business indicators
   - Assemble and support the right team
   - Review customer data
   - Narrow down project focus. Develop project purpose statement

   Typical tools
   Brainstorming, Affinity Diagram, Check Sheet, Control Chart, Histogram, Interrelationship Digraph, Pareto Chart, Prioritization Matrices, Process Capability, Radar Chart, Run Chart

2. **Describe the current process surrounding the improvement opportunity.**
   - Select the relevant process or process segment to define the scope of the project
   - Describe the process under study

   Typical tools
   Brainstorming, Macro, Top-down, and Deployment Flowcharts, Tree Diagram

3. **Describe all of the possible causes of the problem and agree on the root cause(s).**
   - Identify and gather helpful facts and opinions on the cause(s) of the problem
   - Confirm opinions on root cause(s) with data whenever possible

Typical tools
Affinity Diagram, Brainstorming, C&E/Fishbone Diagram, Check Sheet, Force Field Analysis, Interrelationship Digraph, Multivoting, Nominal Group Technique, Pareto Chart, Run Chart, Scatter Diagram

4. **Develop an effective and workable solution and action plan, including targets for improvement.**
   - Define and rank solutions
   - Plan the change process: What? Who? When?
   - Do contingency planning when dealing with new and risky plans
   - Set targets for improvement and establish monitoring methods

Typical tools
Activity Network Diagram, Brainstorming, Flowchart, Gantt Chart, Multivoting, Nominal Group Technique, PDPC, Prioritization Matrices, Matrix Diagram, Tree Diagram

Do

5. **Implement the solution or process change.**
   - It is often recommended to try the solution on a small scale first
   - Follow the plan and monitor the milestones and measures

Typical tools
Activity Network Diagram, Flowchart, Gantt Chart, Matrix Diagram, and other project management methods, as well as gathering ongoing data with Run Charts, Check Sheets, Histograms, Process Capability, and Control Charts.

Check

6. Review and evaluate the result of the change.
- Confirm or establish the means of monitoring the solution. Are the measures valid?
- Is the solution having the intended effect? Any unintended consequences?

Typical tools
Check Sheets, Control Chart, Flowchart, Pareto Chart, Run Chart

Act

7. Reflect and act on learnings.
- Assess the results and problem-solving process and recommend changes
- Continue the improvement process where needed; standardization where possible
- Celebrate success

Typical tools
Affinity Diagram, Brainstorming, Improvement Storyboard, Radar Chart

Appendix C:

The Memory Jogger™

As was mentioned in the introduction to this book, *The Memory Jogger*™ was developed in 1985 in order to fill an important gap in Continuous Improvement literature. Previously, there had been no books widely available that put the Seven Basic QC Tools in the hands of virtually any manager or employee.

Since 1985, many U.S. companies have made tremendous strides toward Total Quality Management (TQM) by going beyond the concept of simply pursuing incremental daily Continuous Improvement. For example, many organizations have now integrated Hoshin Planning/Policy Deployment as a way to focus yearly strategic breakthrough. Likewise, Quality Function Deployment (QFD) and other forms of Cross-Functional Management are now widely used.

The Memory Jogger is included in this book in an attempt to bring all of the tools for Total Quality together in one publication. Hopefully, this reinforces that the Seven Management and Planning Tools supplement rather than replace the very powerful and necessary Seven Basic QC Tools.

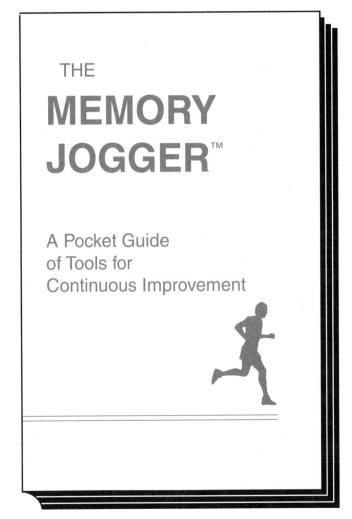

THE

MEMORY JOGGER™

A Pocket Guide of Tools for Continuous Improvement

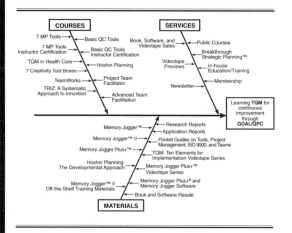

2

GOAL/QPC
is a nonprofit organization striving
to help companies continuously
improve their **Q**uality, **P**roductivity,
and **C**ompetitiveness.

Acknowledgements

The Memory Jogger™ was compiled and edited by Michael Brassard of **GOAL/QPC.** Special thanks to Diane Ritter of **GOAL/QPC** for her special counsel and contribution of statistical materials, and to the following members of the Statistical Resource Committee:

- Ray Caspary — Telesis Systems
- Bryce Colburne — AT&T Technologies, Merrimack Valley
- Gene Fetteroll — Associated Industries of Mass.
- Phil Kendall — Gillette Company
- Ray Lammi — Lammi Associates
- Lawrence LeFebre — GOAL/QPC
- Frank McKernan — Alcoswitch Division of Augat
- Hal Nelson — Alpha Industries

GOAL/QPC also gratefully acknowledges the special contribution of materials by The Kendall Co., Boston, MA; Gould P.C. Division, Andover, MA; Ford Motor Company, Dearborn, MI; Masland, Carlisle, PA; Monsanto Chemical Co., St. Louis, MO; ALCOA, Pittsburgh, PA; and Worcester Memorial Hospital, Worcester, MA.

3

This handbook is designed to help you and every person in your company to IMPROVE DAILY THE PROCEDURES, SYSTEMS, QUALITY, COST, AND YIELDS RELATED TO YOUR JOB. This continuous improvement process is the focus of today's QUALITY REVOLUTION.

In companies that are involved in this revolution, this continuous improvement process has two components:

1) Philosophy
2) Problem-Solving/Graphical Techniques

1. PHILOSOPHY

There are common points in the operating philosophies of these companies. They are as follows:

- Improving quality by removing the causes of problems in the system **inevitably** leads to improved productivity.
- The person doing the job is most knowledgeable about that job.
- People want to be involved and do their jobs well.
- Every person wants to feel like a valued contributor.
- More can be accomplished by working together to improve the system than having individual contributors working around the system.
- A structured problem-solving process using graphical techniques produces better solutions than an unstructured process.
- Graphical problem-solving techniques let you know where you are, where the variations lie, the relative importance of problems to be solved, and whether the changes made have had the desired impact.

- The adversarial relationship between labor and management is counterproductive and outmoded.
- Every organization has undiscovered "gems" waiting to be developed.

2. PROBLEM-SOLVING/GRAPHICAL TECHNIQUES

The rest of this handbook consists of practical descriptions, instructions, and examples of the following techniques:

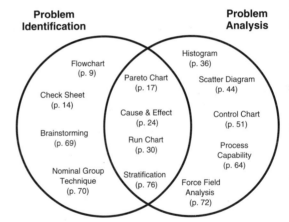

Problem Identification

Problem Analysis

- Flowchart (p. 9)
- Check Sheet (p. 14)
- Brainstorming (p. 69)
- Nominal Group Technique (p. 70)
- Pareto Chart (p. 17)
- Cause & Effect (p. 24)
- Run Chart (p. 30)
- Stratification (p. 76)
- Histogram (p. 36)
- Scatter Diagram (p. 44)
- Control Chart (p. 51)
- Process Capability (p. 64)
- Force Field Analysis (p. 72)

Notice that charts can be used for different purposes in various stages of the problem-solving process. For example, the tools included in the intersecting portion of this **VENN DIAGRAM** can be used in both the Problem Identification and Problem Analysis phase of problem solving.

4

5

TECHNIQUE   SELECTION   GUIDE

Task

Techniques

1. Decide which problem will be addressed first (or next).

- Flowchart (p. 9)
- Check Sheet (p. 14)
- Pareto Chart (p. 17)
- Brainstorming (p. 69)
- Nominal Group Technique (p. 70)

2. Arrive at a statement that describes the problem in terms of what it is specifically, where it occurs, when it happens, and its extent.

- Check Sheet (p. 14)
- Pareto Chart (p. 17)
- Run Chart (p. 30)
- Histogram (p. 36)
- Pie Chart (p. 75)
- Stratification (p. 76)

3. Develop a complete picture of all the possible causes of the problem.

- Check Sheet (p. 14)
- Cause & Effect Diagram (p. 24)
- Brainstorming (p. 69)

4. Agree on the basic cause(s) of the problem.

- Check Sheet (p. 14)
- Pareto Chart (p. 17)
- Scatter Diagram (p. 44)
- Brainstorming (p. 69)
- Nominal Group Technique (p. 70)

5. Develop an effective and implementable solution and action plan.

- Brainstorming (p. 69)
- Force Field Analysis (p. 72)
- Management Presentation (p. 74)
- Pie Chart (p. 75)
- Add'l Bar Graphs (p. 77)

6. Implement the solution and establish needed monitoring procedures and charts.

- Pareto Chart (p. 17)
- Histogram (p. 36)
- Control Chart (p. 51)
- Process Capability (p. 64)
- Stratification (p. 76)

6

7

HOW TO USE *THE MEMORY JOGGER* ™

The Memory Jogger™ is designed as a convenient and quick reference guide on the job. It is therefore organized around symbols and color highlights that are eye-catching and easy to remember. The following is a legend explaining each symbol and its use.

Making Ready

An important step in a problem-solving process is selecting the right tool for the situation. Each section on a new tool will begin with a boxed description that describes when it should be used. Always check this feature first to ensure that the tool meets your needs.

Cruising

Turn to this portion of each section to find construction guidelines. This is the action phase that provides you with step-by-step instructions and helpful formulas. Turn to this feature in each section when you have basic how-to questions.

Finishing The Course

This portion of each section shows each tool in its final form. There are examples from **Manufacturing, Administration/Service, and Daily Life** to display the widespread applications of each tool. Refer to this portion when you need to see the proper form of the finished charting technique.

Caution

The boxed portion at the end of each section describes helpful construction and interpretation tips for that charting technique. Be sure to use this section to avoid making some of the most common mistakes when constructing and analyzing these tools.

8

> **Flowchart:** When you need to identify the actual ideal paths that any product or service follows in order to identify deviations.

FLOWCHART

A Flowchart is a pictorial representation showing all of the steps of a process. Flowcharts provide excellent documentation of a program and can be a useful tool for examining how various steps in a process are related to each other. Flowcharting uses easily recognizable symbols to represent the type of processing performed.

By studying these charts you can often uncover loopholes which are potential sources of trouble. Flowcharts can be applied to anything from the travels of an invoice or the flow of materials, to the steps in making a sale or servicing a product.

The Flowchart is most widely used in problem identification in a process called IMAGINEERING. The people with **important** knowledge about the process meet to:

1.) Draw a Flowchart of what steps the process **actually** follows.
2.) Draw a Flowchart of what steps the process **should** follow if everything worked right.
3.) Compare the two charts to find where they are different because this is where the problems arise.

9

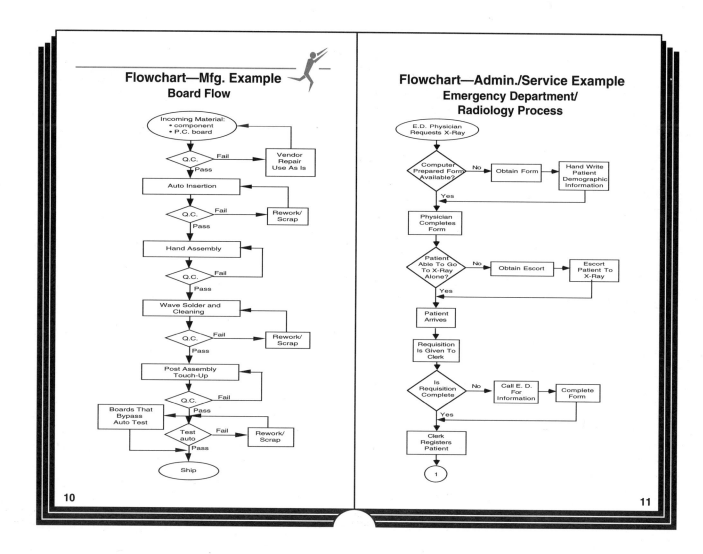

Flowchart—Mfg. Example
Board Flow

Incoming Material:
- component
- P.C. board

Q.C. — Fail → Vendor Repair Use As Is
Pass ↓
Auto Insertion
Q.C. — Fail → Rework/Scrap
Pass ↓
Hand Assembly
Q.C. — Fail
Pass ↓
Wave Solder and Cleaning
Q.C. — Fail → Rework/Scrap
Pass ↓
Post Assembly Touch-Up
Q.C. — Fail
Pass ↓
Boards That Bypass Auto Test
Test auto — Fail → Rework/Scrap
Pass ↓
Ship

Flowchart—Admin./Service Example
Emergency Department/ Radiology Process

E.D. Physician Requests X-Ray

Computer Prepared Form Available? — No → Obtain Form → Hand Write Patient Demographic Information
Yes ↓
Physician Completes Form

Patient Able To Go To X-Ray Alone? — No → Obtain Escort → Escort Patient To X-Ray
Yes ↓
Patient Arrives

Requisition Is Given To Clerk

Is Requisition Complete — No → Call E. D. For Information → Complete Form
Yes ↓
Clerk Registers Patient

1

10

11

Flowchart—Daily Example
Turning On A Television

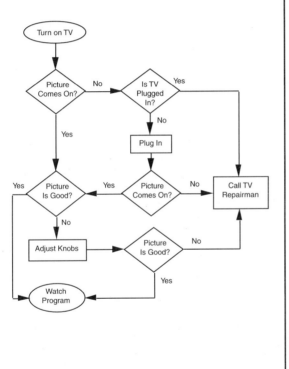

 Construction/Interpretation Tips
Flowchart

- Define the boundaries of the process clearly.
- Use the simplest symbols possible.
- Make sure every feedback loop has an escape.
- There is usually only one output arrow out of a *process* box. Otherwise, it may require a *decision* diamond.

12

13

Check Sheet: When you need to gather data based on sample observations in order to begin to detect patterns. This is the logical point to start in most problem-solving cycles.

CHECK SHEET

Check Sheets are simply an easy to understand form used to answer the question, "How often are certain events happening?" It starts the process of translating "opinions" into "facts." Constructing a Check Sheet involves the following steps:

1.) Agree on the event being observed. Everyone has to be looking for the same thing.
2.) Decide on a time period to collect data. This could range from hours to weeks.
3.) Design a form that is clear and easy to use, with columns clearly labeled and adequate space for entering data.
4.) Collect the data consistently and honestly. Make sure there is time allowed for this data-gathering task.

Problem	Month			
	1	2	3	Total
A	II	II	I	5
B	I	I	I	3
C	ⅬⅡⅠ	II	ⅬⅡⅠ	12
Total	8	5	7	20

Check Sheet—Mfg. Example
Bearing Defects

Defect	May				Total
	6	7	8	9	
Wrong Size	ⅬⅡⅠ I	ⅬⅡⅠ	ⅬⅡⅠ III	ⅬⅡⅠ II	26
Wrong Shape	I	III	III	II	9
Wrong Dept.	ⅬⅡⅠ	I	I	I	8
Wrong Weight	ⅬⅡⅠ ⅬⅡⅠ ⅬⅡⅠ	ⅬⅡⅠ ⅬⅡⅠ	ⅬⅡⅠ ⅬⅡⅠ II	ⅬⅡⅠ ⅬⅡⅠ ⅬⅡⅠ	52
Wrong Smoothness	II	III	I	I	7
Total	29	22	25	26	102

Check Sheet—Admin./Service Ex.
Typing Mistakes In Department A

Mistakes	March			Total
	1	2	3	
Centering	II	III	III	8
Spelling	ⅬⅡⅠ II	ⅬⅡⅠ ⅬⅡⅠ I	ⅬⅡⅠ	23
Punctuation	ⅬⅡⅠ ⅬⅡⅠ ⅬⅡⅠ ⅬⅡⅠ	ⅬⅡⅠ ⅬⅡⅠ	ⅬⅡⅠ ⅬⅡⅠ ⅬⅡⅠ	40
Missed Paragraph	II	I	I	4
Wrong Numbers	III	IIII	III	10
Wrong Page #'s	I	I	II	4
Tables	IIII	ⅬⅡⅠ	IIII	13
Total	34	35	33	102

14

15

Check Sheet—Daily Example
Reasons For Disagreements

Reason	Day					Total
	M	T	W	T	F	
Money	卌	II	I	卌	卌 II	20
Sex	II	II	II	II	II	10
Children	卌	II	卌 II	I	IIII	19
Total	12	6	10	8	13	49

 Construction/Interpretation Tips Check Sheet

- Make sure that observations/samples are as representative as possible.
- Make sure the sampling process is efficient so that people have time to do it.
- In doing a Check Sheet, you may not be aware that the population (universe) being sampled is non-homogeneous (not from the same machine, person, etc.). The population must be homogeneous. If not, it must first be stratified (grouped), with each grouping sampled individually.

Pareto Chart: When you need to display the relative importance of all of the problems or conditions in order to: choose the starting point for problem solving, monitor success, or identify the basic cause of a problem.

PARETO CHART

A Pareto Chart is a special form of vertical bar graph which helps us to determine which problems to solve in what order. Doing a Pareto Chart based upon either Check Sheets or other forms of data collection helps us direct our attention and efforts to the truly important problems. We will generally gain more by working on the tallest bar than tackling the smaller bars.

Pareto Chart
Defects Found At In-Process Inspection

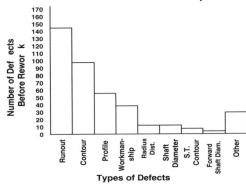

16

17

STEPS IN CONSTRUCTING A PARETO CHART

1.) Select the problems that are to be compared and rank-ordered by:
   a.) **Brainstorming,** e.g., "What are our major quality problems in Department A?"
   b.) **Using Existing Data,** e.g., "Let's take a look at Department A's quality reports over the last month to find the major problem areas."

2.) Select the standard for comparison unit of measurement, e.g., annual cost, frequency, etc.

3.) Select time period to be studied, e.g., 8 hours, 5 days, 4 weeks.

4.) Gather necessary data on the site of each category, e.g., "Defect A occurred X times in the last 6 months" or "Defect B cost X dollars in the last 6 months."

5.) Compare the frequency or cost of each category relative to all other categories, e.g., "Defect A happened 75 times; Defect B happened 107 times; Defect C happened 35 times," or "Defect A cost $750 annually; Defect B cost $535 annually."

6.) List the categories from left to right on the horizontal axis in their order of decreasing frequency or cost. The categories containing the fewest items can be combined into an "other" category, which is placed on the extreme right as the last bar.

7.) Above each classification or category, draw a rectangle whose height represents the frequency or cost in that classification.*

18

* Additional Features of Pareto Charts:

• Often the "raw data" is recorded on the left vertical axis with a percentage scale on the right vertical axis. Make sure that the two axes are drawn to scale, e.g., 100% is opposite the total frequency or cost; 50% is opposite the halfway point in the raw data.

• From the top of the tallest bar and moving upward from left to right, a line can be added that shows the cumulative frequency of the categories. This answers such questions as, "How much of the total is accounted for by the first three categories?"

19

DIFFERENT USES OF A PARETO CHART

1.) To identify the most important problems through the use of different measurement scales, e.g., frequency, cost.

Lesson: The most frequent problems are not always the most costly.

Field Service Customer Complaints:

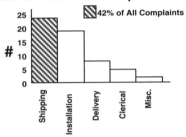

Cost To Rectify Field Service Complaints:

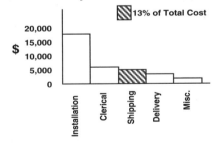

2.) To analyze different groupings of data, e.g., by product, by machine, by shift.

Lesson: If clear differences don't emerge, regroup the data. Use your imagination.

Pareto Analysis Of The Number Of Defects

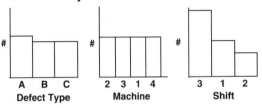

3.) To measure the impact of changes made in a process, e.g., before and after comparisons.

Lesson: You don't know how much better you are if you don't know where you were before the change.

Component Defects (Station #2)

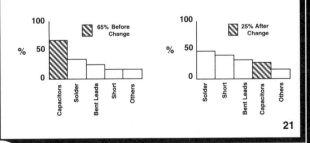

20

21

4.) To break down broad causes into more and more specific parts.

Lesson: Cure the cause, not the symptom.

Types of Injury

Causes of Eye Injury

⚠️ **Construction/Interpretation Tips**
Pareto Chart

- Use common sense—two key customer complaints may deserve more attention than 100 other complaints, depending on who the customer is and what the complaint is.
  Mark chart clearly to show standard of
- measurement ($, %, or #).

> **Cause & Effect Diagram:** When you need to identify and explore and display the possible causes of a specific problem or condition.

CAUSE & EFFECT DIAGRAM

The Cause & Effect Diagram was developed to represent the relationship between some "effect" and all the possible "causes" influencing it. The effect or problem is stated on the right side of the chart and the major influences or "causes" are listed to the left.

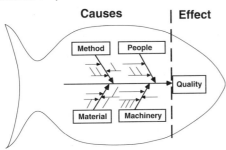

(Fishbone Diagram)

Cause & Effect Diagrams are drawn to clearly illustrate the various causes affecting a process by sorting out and relating the causes. For every effect there are likely to be several major categories of causes. The major causes might be summarized under four categories, People, Machines, Methods, and Materials. In administrative areas it may be more helpful to use the 4P's: Policies, Procedures, People, and Plant. Remember that these categories are only suggestions. You may use any major category that emerges or helps people think creatively.

A well-detailed Cause & Effect Diagram will take on the shape of fishbones and hence the alternate name Fishbone Diagram. From this well-defined list of possible causes, the most likely are identified and selected for further analysis. When examining each cause look for things that have changed, deviations from the norm or patterns. Remember, look to cure the cause and not the symptoms of the problem. Push the causes back as much as is practically possible.

STEPS IN CONSTRUCTING A CAUSE & EFFECT DIAGRAM

1.) Generate the causes needed to build a Cause & Effect Diagram in one of two ways:
    a.) Structured **Brainstorming** about possible causes without previous preparation.
    b.) Ask members of the team to spend time between meetings using simple **Check Sheets** to track possible causes and to examine the production process steps closely.
2.) Construct the actual Cause & Effect Diagram by:
    a.) Placing problem statement in box on the right.
    b.) Drawing the traditional major cause category steps in the production process, or any causes that are helpful in organizing the most important factors.
    c.) Placing the **Brainstormed** ideas in the appropriate major categories.
    d.) For each cause ask, "Why does it happen?" and list responses as branches off the major causes.
3.) Interpretation
In order to find the most basic causes of the problem:
    a.) Look for causes that appear repeatedly.
    b.) Reach a team consensus.
    c.) Gather data to determine the relative frequencies of the different causes.

24

25

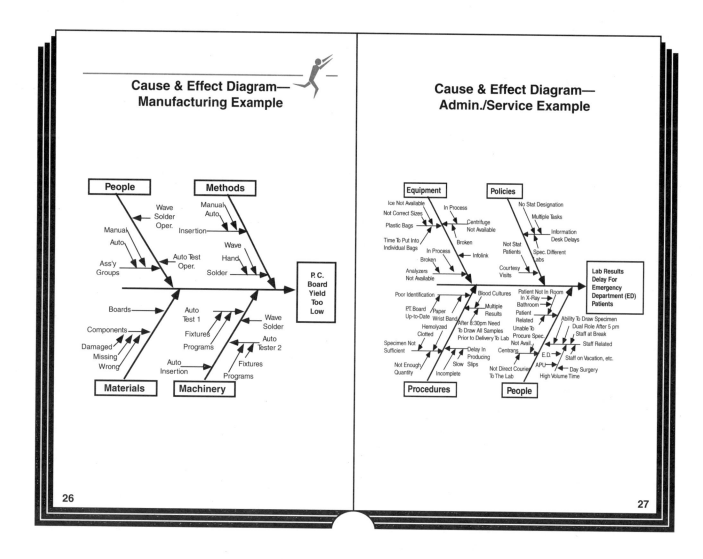

Cause & Effect Diagram—Manufacturing Example

People

Wave Solder Oper.

Manual
Auto

Ass'y Groups

Auto Test Oper.

Methods

Manual
Auto
Insertion

Wave
Hand
Solder

Boards

Components

Damaged
Missing
Wrong

Auto Insertion

Auto Test 1

Fixtures

Programs

Wave Solder

Auto Tester 2

Fixtures

Programs

Materials   **Machinery**

P. C. Board Yield Too Low

Cause & Effect Diagram—Admin./Service Example

Equipment

Ice Not Available
Not Correct Sizes
Plastic Bags
Time To Put Into Individual Bags

In Process
Centrifuge Not Available
Broken
In Process
Infolink
Broken

Analyzers Not Available

Policies

No Stat Designation
Multiple Tasks
Information Desk Delays
Spec. Different Labs

Not Stat Patients

Courtesy Visits

Poor Identification
P.T. Board Up-to-Date
Paper Wrist Band
Hemolyzed Clotted
Specimen Not Sufficient
Not Enough Quantity

Blood Cultures
Multiple Results
After 8:30pm Need To Draw All Samples Prior to Delivery To Lab
Delay In Producing Slips
Slow
Incomplete

Patient Not In Room
In X-Ray
Bathroom
Patient Related
Unable To Procure Spec.
Not Avail.
Centrans
Not Direct Courier To The Lab

Ability To Draw Specimen
Dual Role After 5 pm
Staff at Break
Staff Related
E.D.
APU
Staff on Vacation, etc.
Day Surgery
High Volume Time

Lab Results Delay For Emergency Department (ED) Patients

Procedures   **People**

26

27

Cause & Effect Diagram—Daily Example

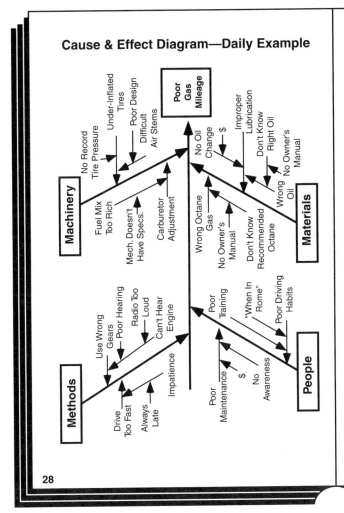

Construction/Interpretation Tips
Cause & Effect Diagram

- Try not to go far beyond the area of control of the group in order to minimize frustration.
- If ideas are slow in coming use the major cause categories as catalysts, e.g., "What in *materials* is causing...?"
- Use as few words as possible.
- Make sure everyone agrees completely on the problem statement.
- The most widely used type of Cause & Effect Diagram is the **Dispersion Analysis** type which is the type shown in *The Memory Jogger*™. It is constructed by placing individual causes within each "major" cause category and asking the following question of each item, "Why does this cause (dispersion) happen?" Other common types of Cause & Effect Diagrams are as follows:

  a. **Process Classification Cause & Effect Diagrams** sequentially list all the steps in a process. The same cause category arrows as in the Dispersion Analysis type branch off the line between each process step. The same questions are then applied to each cause category as in the Dispersion Analysis type diagram.

  b. **Cause Enumeration Cause & Effect Diagrams** are almost identical to the Dispersion Analysis type. The only real difference rests in the fact that Cause Enumeration first organizes all the possible causes in list form and then places them in the major cause categories.

28

29

Run Chart: When you need to do the simplest possible display of trends within observation points over a specified time period.

RUN CHART

Run Charts are employed to visually represent data. They are used to monitor a process to see whether or not the long range average is changing.

Run Charts are the simplest tool to construct and use. Points are plotted on the graph in the order in which they become available. It is common to graph the results of a process such as machine downtime, yield, scrap, typographical errors, or productivity as they vary over time.

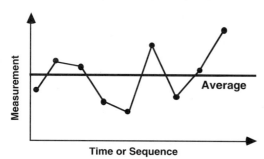

A danger in using a Run Chart is the tendency to see every variation in data as being important. The Run Chart, like the other charting techniques, should be **used to focus attention on truly vital changes in the process.**

30

One of the most valuable uses of Run Charts is to identify meaningful trends or shifts in the average. For example when monitoring any process, it is expected that we should find an equal number of points falling above and below the average. Therefore, when nine points "run" on one side of the average it indicates a statistically unusual event and that the average has changed. Such changes should always be investigated. If the shift is favorable, it should be made a permanent part of the system. If it is unfavorable, it should be eliminated.

An alternate type of pattern that can occur is a trend of six or more points steadily increasing or decreasing with no reversals. Neither pattern would be expected to happen based on random chance. Thus, this would likely indicate an important change and the need to investigate.

31

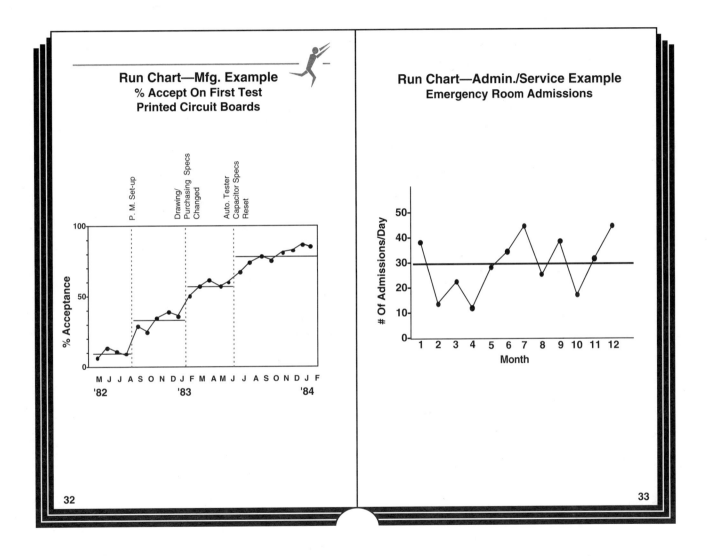

Run Chart—Mfg. Example
% Accept On First Test
Printed Circuit Boards

P. M. Set-up

Drawing/
Purchasing Specs
Changed

Auto. Tester
Capacitor Specs
Reset

% Acceptance

100

50

0

M J J A S O N D J F M A M J J A S O N D J F
'82 '83 '84

Run Chart—Admin./Service Example
Emergency Room Admissions

Of Admissions/Day

50
40
30
20
10
0

1 2 3 4 5 6 7 8 9 10 11 12
Month

32

33

Run Chart—Daily Example
Family Expenditure/Month

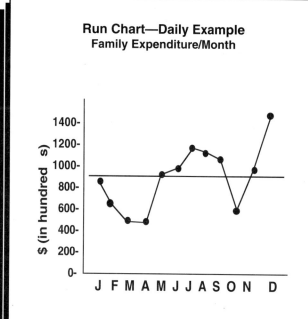

Construction/Interpretation Tips
Run Chart

- The *y axis* is the vertical side of the graph.
- The *x axis* is the horizontal side of the graph.
- A marked point indicates the measurement or quantity observed or sampled at one point in time.
- Data points should be connected for easy use and interpretation.
- The time period covered and unit of measurement used must be clearly marked.
- Collected data must be kept in the order that it was gathered. Since it is tracking a characteristic over time, the sequence of data points is critical.

34

35

> **Histogram:** When you need to discover and display the distribution of data by bar graphing the number of units in each category.

HISTOGRAM

As we have already seen with the **Pareto Chart**, it is very helpful to display in bar graph form the frequency with which certain events occur (frequency distribution). The **Pareto Chart**, however, only deals with characteristics of a product or service, e.g., type of defect, problem, safety hazards (attribute data). A Histogram takes measurement data, e.g., temperature, dimensions, and displays its distribution. This is critical since we know that all repeated events will produce results that vary over time. A Histogram reveals the amount of variation that any process has within it. A typical Histogram would look like this:

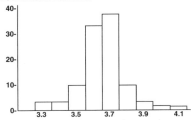

The Histogram pictured here shows the greatest number of units are at the center measurement with roughly an equal number of units falling on either side of that point. Many repeated samples of data under statistical control follow this pattern. Other data displays patterns

with the data "piled up" at points away from the center. Such a distribution is referred to as being "skewed." The important thing to remember is that you are looking for surprises such as distributions that should be naturally "normal" but are not. The same is true for predictably skewed distributions. In addition to the shape of the distribution you are also looking for:

a.) Whether the "spread" of the curve falls within specifications. If not, how much falls outside of specifications. (**VARIABILITY**)

b.) Whether the curve is centered at the right place. Are most items on the "high or low side?" (**SKEWNESS**)

Illustrations Of Variability

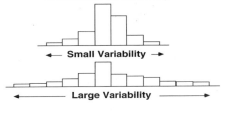

Illustrations Of Skewness

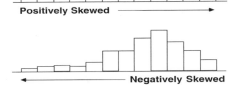

STEPS IN CONSTRUCTING A HISTOGRAM

There will be more detailed instructions for constructing Histograms than most of the other tools. This is being stressed because of the confusion that seems to be created when deciding on the number of classes (bars), the class boundaries, etc.

Start with an unorganized set of numbers such as the following:

```
 9.9   9.3 10.2   9.4 10.1   9.6   9.9 10.1   9.8
 9.8   9.8 10.1   9.9   9.7   9.8   9.9 10.0   9.6
 9.7   9.4   9.6 10.0   9.8   9.9 10.1 10.4 10.0
10.2 10.1   9.8 10.1 10.3 10.0 10.2   9.8 10.7
 9.9 10.7   9.3 10.3   9.9   9.8 10.3   9.5   9.9
 9.3 10.2   9.2   9.9   9.7   9.9   9.8   9.5   9.4
 9.0   9.5   9.7   9.7   9.8   9.8   9.3   9.6   9.7
10.0   9.7   9.4   9.8   9.4   9.6 10.0 10.3   9.8
 9.5   9.7 10.6   9.5 10.1 10.0   9.8 10.1   9.6
 9.6   9.4 10.1   9.5 10.1 10.2   9.8   9.5   9.3
10.3   9.6   9.7   9.7 10.1   9.8   9.7 10.0 10.0
 9.5   9.5   9.8   9.9   9.2 10.0 10.0   9.7   9.7
 9.9 10.4   9.3   9.6 10.2   9.7   9.7   9.7 10.7
 9.9 10.2   9.8   9.3   9.6   9.5   9.6 10.7
```

These numbers refer to the thickness of a certain key material in a process.

STEP 1: Count the number of data points in the set data. For our example above there are 125 data points (n=125).

STEP 2: Determine the range, R, for the entire data set. The range is the smallest value in the set of data subtracted from the largest value. In our case, the range is equal to 10.7 minus 9.0. Thus, the range equals 1.7.

STEP 3: Divide the data set* into a certain number of classes, referred to as K. The table below provides an approximate guideline for dividing your set of data into a reasonable number of classes. For our example, 125 data points would be broken down into 7-12 classes. We will use K=10 classes.

Number Of Data Points	Number Of Classes (K)
Under 50	5-7
50-100	6-10
100-250	7-12
Over 250	10-20

STEP 4: Determine the class width, H. A convenient formula is as follows:

$$H = \frac{R}{K} = \frac{1.7}{10} = .17$$

In this case, as in most, it helps to round off H but carry the number to one more decimal point than in the original data set. For our purposes, 0.20 would appear appropriate.

* In previous printings this was referred to as "range value." It has been revised for greater clarity.

STEP 5: Determine the class boundary, or end points. For simple determination of the class boundaries, take the smallest individual measurement in the data set. Use this number or round to the next appropriate lower number. This will be the lower end point for our first class boundary. In our example this would be 9.00. Now take this number and add the class width to it, 9.00 + 0.20 = 9.20. Thus, the next lower class boundary would begin at 9.20. The first class would be 9.00 and everything up to, **but not including** 9.20, 9.00 through 9.19! The second class would begin at 9.20 and be everything up to, but not including 9.40! This makes each class mutually exclusive, that is each of our data set points will fit into one, **and only one**, class. Finally, consecutively add the class width, 0.20, to the lowest class boundary until the correct number of classes, approximately 10 and containing the range of all our numbers, is obtained.

STEP 6: Construct a frequency table based on the values computed above (i.e., number of classes, class width, class boundary). The frequency table is actually a Histogram in a tabular form. A frequency table based on the thickness data is shown below:

Class #	Class Boundaries	Mid-point	Frequency	Total
1	9.00-9.19	9.1	I	1
2	9.20-9.39	9.3	ЖНТ IIII	9
3	9.40-9.59	9.5	ЖНТ ЖНТ ЖНТ I	16
4	9.60-9.79	9.7	ЖНТ ЖНТ ЖНТ ЖНТ ЖНТ ЖНТ II	27
5	9.80-9.99	9.9	ЖНТ ЖНТ ЖНТ ЖНТ ЖНТ ЖНТ I	31
6	10.00-10.19	10.1	ЖНТ ЖНТ ЖНТ ЖНТ II	22
7	10.20-10.39	10.3	ЖНТ ЖНТ II	12
8	10.40-10.59	10.5	II	2
9	10.60-10.79	10.7	ЖНТ	5
10	10.80-10.99	10.9		0

STEP 7: Construct the Histogram based on the frequency table. A Histogram is a graphical picture of a frequency table. It provides us with a quick picture of the distribution for the measured characteristic. A Histogram for our example is shown below:

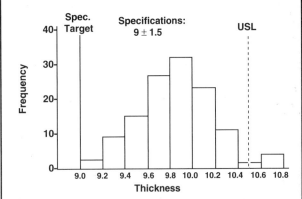

As pointed out earlier, the Histogram is an important diagnostic tool because it gives a "bird's-eye view" of the variation in a data set. In our case, the data appears to have a central tendency around 9.8 to 9.99. It also appears the data creates close to a normal curve. The specification for the thickness characteristic is 7.5 to 10.5, with a target of 9. Thus, we can see that our Histogram indicates the process is running high and that defective material is being made.

40

41

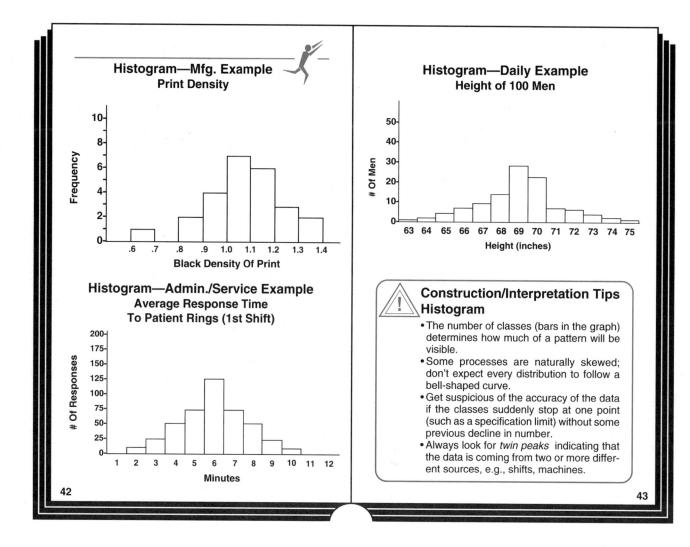

Histogram—Mfg. Example
Print Density

Frequency axis (y): 0, 2, 4, 6, 8, 10
Black Density Of Print (x): .6 .7 .8 .9 1.0 1.1 1.2 1.3 1.4

Histogram—Admin./Service Example
**Average Response Time
To Patient Rings (1st Shift)**

Of Responses (y): 0, 25, 50, 75, 100, 125, 150, 175, 200
Minutes (x): 1 2 3 4 5 6 7 8 9 10 11 12

42

Histogram—Daily Example
Height of 100 Men

Of Men (y): 0, 10, 20, 30, 40, 50
Height (inches) (x): 63 64 65 66 67 68 69 70 71 72 73 74 75

⚠ Construction/Interpretation Tips
Histogram

- The number of classes (bars in the graph) determines how much of a pattern will be visible.
- Some processes are naturally skewed; don't expect every distribution to follow a bell-shaped curve.
- Get suspicious of the accuracy of the data if the classes suddenly stop at one point (such as a specification limit) without some previous decline in number.
- Always look for *twin peaks* indicating that the data is coming from two or more different sources, e.g., shifts, machines.

43

> **Scatter Diagram:** When you need to display what happens to one variable when another variable changes in order to test a theory that the two variables are related.

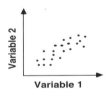

SCATTER DIAGRAM

A Scatter Diagram is used to study the possible relationship between one variable and another. The Scatter Diagram is used to test for possible **cause** and **effect** relationships. It cannot prove that one variable **causes** the other, but it does make it clear whether a relationship exists and the strength of that relationship.

A Scatter Diagram has a horizontal axis (x-axis), to represent the measurement values of one variable, and a vertical axis (y-axis), to represent the measurements of the second variable. A "typical" Scatter Diagram may look like this:

Notice how the plotted points form a clustered pattern. The direction and "tightness" of the cluster give you a clue as to the strength of the relationship between variable 1 and variable 2. The more that this cluster resembles a straight line, the stronger the correlation between the variables. This makes sense since a straight line would mean that every time one variable would change the other would change by the same amount.

44

1.) Collect 50 to 100 paired samples of data that you think may be related and construct a data sheet as follows:

Person	Weight	Height
1	160 lbs.	70 inches
2	180 "	61 "
3	220 "	75 "
.	.	.
.	.	.
50	105 "	61 "

2.) Draw the horizontal and vertical axes of the diagram. The values should get higher as you move up and to the right on each axis. The variable that's being investigated as the possible "cause" is usually on the horizontal and the "effect" variable is usually on the vertical.

3.) Plot the data on the diagram. If you find the values being repeated, circle that point as many times as appropriate. The resulting diagram may look like this:

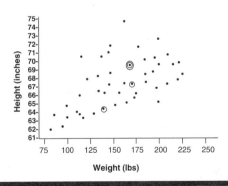

45

The following are the various patterns and meanings that Scatter Diagrams can have:

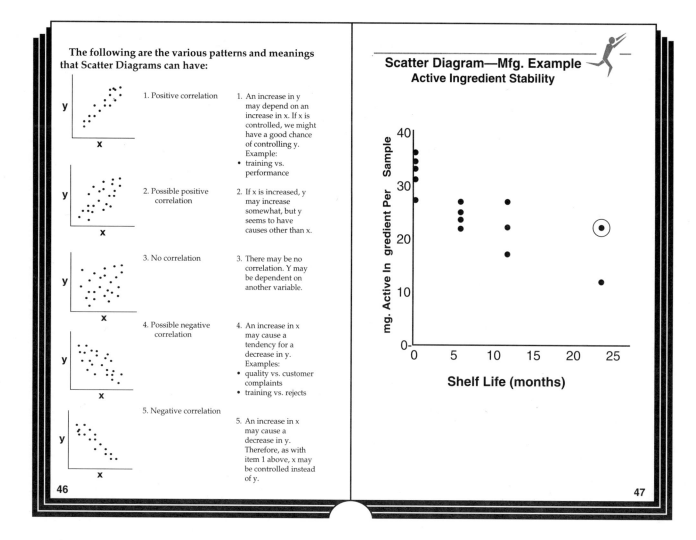

1. Positive correlation

1. An increase in y may depend on an increase in x. If x is controlled, we might have a good chance of controlling y. Example:
   • training vs. performance

2. Possible positive correlation

2. If x is increased, y may increase somewhat, but y seems to have causes other than x.

3. No correlation

3. There may be no correlation. Y may be dependent on another variable.

4. Possible negative correlation

4. An increase in x may cause a tendency for a decrease in y. Examples:
   • quality vs. customer complaints
   • training vs. rejects

5. Negative correlation

5. An increase in x may cause a decrease in y. Therefore, as with item 1 above, x may be controlled instead of y.

46

Scatter Diagram—Mfg. Example
Active Ingredient Stability

47

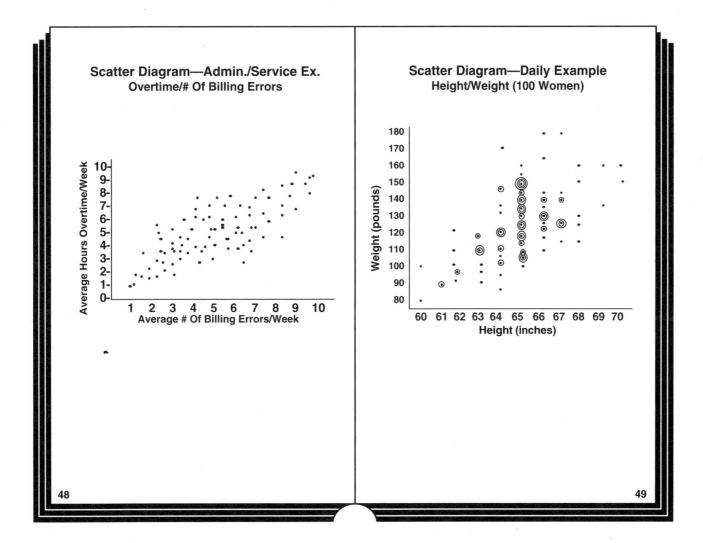

Scatter Diagram—Admin./Service Ex.
Overtime/# Of Billing Errors

Average Hours Overtime/Week vs *Average # Of Billing Errors/Week*

Scatter Diagram—Daily Example
Height/Weight (100 Women)

Weight (pounds) vs *Height (inches)*

48

49

Construction/Interpretation Tips
Scatter Diagram

- A negative relationship (as y increases, x decreases) is as important as a positive relationship (as x increases, y increases).
- You can only say that x and y are related and not that one *causes* the other.
- The examples in this section were based on straight line correlations: y=a+bx. However, this is not the only form of relationship that can be routinely encountered: $y=e^x$, $y=x^2$, and $y^2=x$ are just a few of the many types that can occur.
- There are statistical tests available to test the exact degree of correlation but are beyond the scope of this book.

Control Chart: When you need to discover how much variability in a process is due to random variation and how much is due to unique events/individual actions in order to determine whether a process is in statistical control.

CONTROL CHART

A Control Chart is simply a run chart with **statistically** determined upper (Upper Control Limit) and possibly lower (Lower Control Limit) lines drawn on either side of the process average.

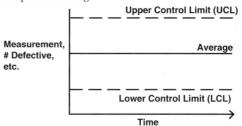

These limits are calculated by running a process untouched (i.e., according to standard procedures but without any extra "tweaking" adjustments), taking samples, and plugging the sample averages into the appropriate formula. You can now plot the sample averages onto a chart to determine whether any of the points fall between or outside of the limits or form unlikely patterns. If either of these happen, the process is said to be "out of control."

The fluctuation of the points within the limits results from variation built into the process. This results from **common causes** within the **system** (e.g., design, choice of

50

51

machine, preventive maintenance), and can only be affected by changing that system. However, points outside of the limits come from a **special cause** (e.g., people errors, unplanned events, freak occurrences), that is not part of the way that the process normally operates, or from an unlikely combination of process steps. These special causes must be eliminated before the Control Chart can be used as a monitoring tool. Once this is done, the process would be "in control" and samples can be taken at regular intervals to make sure that the process doesn't fundamentally change.

REMEMBER: "Control" doesn't necessarily mean that the product or service will meet your needs. It only means that the process is **consistent** (may be consistently bad). For example:

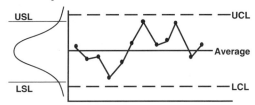

In this case, the process is in control but it is not capable of meeting the specification. The curve to the left of the Control Chart shows that the specification limits are narrower than the Control Charted process. Either you improve the process or you change the specifications. Just remember that specifications are what you think you need and control limits are what the process can do consistently.

STEPS IN CONSTRUCTION OF CONTROL CHARTS AND CRITICAL FORMULAS

Variables Control Chart:
When samples are expressed in **quantitative** units of measurement, e.g., length, weight, time.

$\overline{X}$—R Chart =
Plotting the Average & Range of Data Collected

Calculate the Average ($\overline{X}$) and Range (R) of each subgroup:

$$\overline{X} = \frac{X_1 + X_2 + \dots + X_n}{n} \qquad n = \text{\# of samples}$$

$$R = X_{max} - X_{min}$$

Calculate the Average Range ($\overline{R}$) and the Process Average ($\overline{\overline{X}}$):

$$\overline{\overline{X}} = \frac{\overline{X}_1 + \overline{X}_2 + \dots + \overline{X}_k}{k} \qquad k = \text{\# of subgroups}$$

$$\overline{R} = \frac{R_1 + R_2 + \dots + R_k}{k} \qquad \text{(20-25 groups)}$$

Calculate the Control Limits:

$$UCL\overline{x} = \overline{\overline{X}} + A_2\overline{R} \qquad LCL\overline{x} = \overline{\overline{X}} - A_2\overline{R}$$
$$UCL_R = D_4\overline{R} \qquad LCL_R = D_3\overline{R}$$

Table of Factors for $\overline{X}$ & R Charts

Number of observations in subgroup (n)	Factors for X Chart A_2	Factors for R Chart Lower D_3	Factors for R Chart Upper D_4
2	1.880	0	3.268
3	1.023	0	2.574
4	0.729	0	2.282
5	0.577	0	2.114
6	0.483	0	2.004
7	0.419	0.076	1.924
8	0.373	0.136	1.864
9	0.337	0.184	1.816
10	0.308	0.223	1.777

STEPS IN CONSTRUCTION OF CONTROL CHARTS AND CRITICAL FORMULAS

Attributes Control Chart:
When sample reflects **qualitative** characteristics, e.g., is/is not defective, go/no go.

The p Chart = Proportion Defective

$p = \dfrac{\text{number of rejects in subgroups}}{\text{number inspected in subgroup}}$

$\bar{p} = \dfrac{\text{total number rejects}}{\text{total number inspected}}$

$UCL_p{}^* = \bar{p} + 3\dfrac{\sqrt{\bar{p}\,(1-\bar{p})}}{\sqrt{n}} \qquad LCL_p{}^* = \bar{p} - 3\dfrac{\sqrt{\bar{p}\,(1-\bar{p})}}{\sqrt{n}}$

The np Chart = Number Defective

$UCL_{np} = n\bar{p} + 3\sqrt{n\bar{p}\,(1-\bar{p})} \qquad LCL_{np} = n\bar{p} - 3\sqrt{n\bar{p}\,(1-\bar{p})}$

The c Chart = Number nonconformities with a constant sample size

$\bar{c} = \dfrac{\text{total nonconformities}}{\text{number of subgroups}}$

$UCL_c = \bar{c} + 3\sqrt{\bar{c}} \qquad LCL_c = \bar{c} - 3\sqrt{\bar{c}}$

The u Chart = Number nonconformities with varying sample size

$\bar{u} = \dfrac{\text{total nonconformities}}{\text{total units inspected}}$

$UCL_u{}^* = \bar{u} + 3\dfrac{\sqrt{\bar{u}}}{\sqrt{n}} \qquad LCL_u{}^* = \bar{u} - 3\dfrac{\sqrt{\bar{u}}}{\sqrt{n}}$

* This formula creates changing control limits. To avoid this, use Average Sample Sizes $\sqrt{\bar{n}}$ for those samples that are ±20% of the average sample size. Calculate individual limits for the samples exceeding ±20%.

INTERPRETING CONTROL CHARTS

The process is said to be "out of control" if:

1.) One or more points fall outside of the control limits or:

2.) When you divide the control chart into zones as follows:

— — — — — — —	**Upper Control Limit (UCL)**
Zone A	
Zone B	
Zone C	
Zone C	**Centerline/Average**
Zone B	
Zone A	
— — — — — — —	**Lower Control Limit (LCL)**

You should take note and examine what has changed and **possibly** make a process adjustment if:

a.) Two points, out of three successive points, are on the same side of the centerline in Zone A or beyond.

b.) Four points, out of five successive points, are on the same side of the centerline in Zone B or beyond.

c.) Nine successive points are on one side of the centerline.

d.) There are six consecutive points, increasing or decreasing.

e.) There are fourteen points, in a row alternating up and down.

f.) There are fifteen points in a row within Zone C (above and below centerline).

(see diagram page 56)

54

55

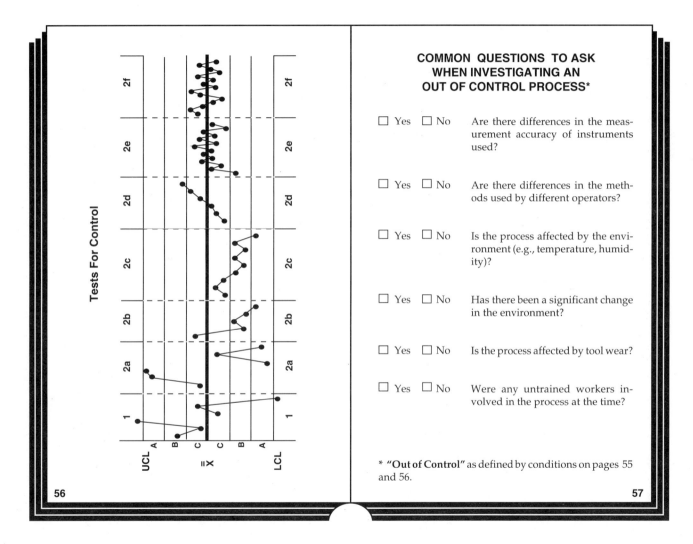

Tests For Control

56

COMMON QUESTIONS TO ASK WHEN INVESTIGATING AN OUT OF CONTROL PROCESS*

☐ Yes ☐ No   Are there differences in the measurement accuracy of instruments used?

☐ Yes ☐ No   Are there differences in the methods used by different operators?

☐ Yes ☐ No   Is the process affected by the environment (e.g., temperature, humidity)?

☐ Yes ☐ No   Has there been a significant change in the environment?

☐ Yes ☐ No   Is the process affected by tool wear?

☐ Yes ☐ No   Were any untrained workers involved in the process at the time?

* **"Out of Control"** as defined by conditions on pages 55 and 56.

57

COMMON QUESTIONS TO ASK WHEN INVESTIGATING AN OUT OF CONTROL PROCESS*

(Continued)

☐ Yes ☐ No   Has there been a change in the source for raw materials?

☐ Yes ☐ No   Is the process affected by operator fatigue?

☐ Yes ☐ No   Has there been a change in maintenance procedures?

☐ Yes ☐ No   Is the machine being adjusted frequently?

☐ Yes ☐ No   Did the samples come from different machines? Shifts? Operators?

☐ Yes ☐ No   Are operators afraid to report "bad news"?

58

Control Chart—Mfg. Example
X̄ & R Chart

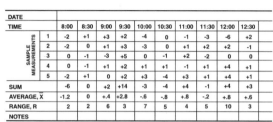

DATE										
TIME	8:00	8:30	9:00	9:30	10:00	10:30	11:00	11:30	12:00	12:30
SAMPLE MEASUREMENTS 1	-2	+1	+3	+2	-4	0	-1	-3	-6	+2
2	-2	0	+1	+3	-3	0	+1	+2	+2	-1
3	0	-1	-3	+5	0	-1	+2	-2	0	0
4	0	-1	+1	+2	+1	+1	-1	+1	+4	+1
5	-2	+1	0	+2	+3	-4	+3	+1	+4	+1
SUM	-6	0	+2	+14	-3	-4	+4	-1	+4	+3
AVERAGE, X̄	-1.2	0	+.4	+2.8	-.6	-.8	+.8	-.2	+.8	+.6
RANGE, R	2	2	6	3	7	5	4	5	10	3
NOTES										

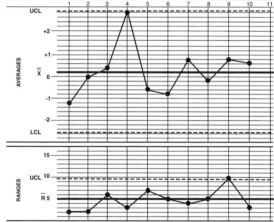

59

Control Chart—Admin./Service Ex.
np Chart
Operating Room Delays/Day

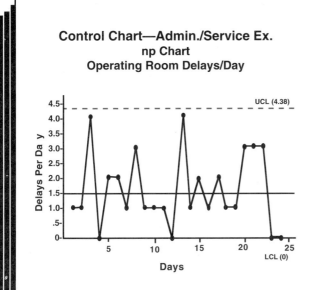

Control Chart—Daily Example
Commuting Times (min.)—A.M.

STEP 1

	Week									
	1	**2**	**3**	**4**	**5**	**6**	**7**	**8**	**9**	**10**
	55	90	100	70	55	75	120	65	70	100
	75	95	75	110	65	85	110	65	85	80
	65	60	75	65	95	65	65	90	60	65
	80	60	65	60	70	65	85	90	65	60
	80	55	65	60	70	65	70	60	75	80
$\overline{X}=$	71	72	76	73	71	71	90	74	71	77
R=	25	40	35	50	40	20	55	30	25	40

STEP 2

$$\overline{\overline{X}} = 74.6$$
$$\overline{R} = 36.0$$
$$n = 5$$
$$k = 10$$

STEP 3

$$UCL\overline{x} = \overline{\overline{X}} + A_2\overline{R}$$
$$= 74.6 + (.58)(36.0)$$
$$= 74.6 + 20.88$$
$$= 95.48$$
$$LCL\overline{x} = \overline{\overline{X}} - A_2\overline{R}$$
$$= 74.6 - 20.88$$
$$= 53.72$$
$$UCL_R = D_4\overline{R}$$
$$= (2.11)(36.0)$$
$$= 75.96$$
$$LCL_R = D_3\overline{R}$$
$$= 0$$

60

61

Commuting Times—A.M.

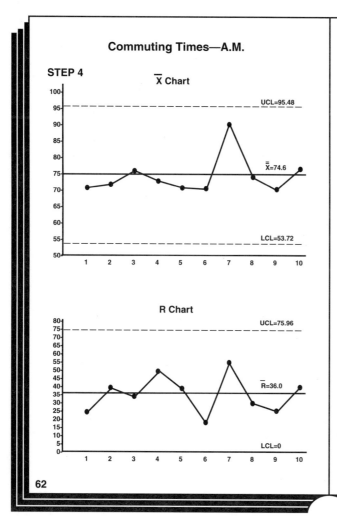

STEP 4

X̄ Chart

UCL=95.48

X̿=74.6

LCL=53.72

R Chart

UCL=75.96

R̄=36.0

LCL=0

Construction/Interpretation Tips
Control Chart

- Generally, collect 20-25 groups of samples before calculating the control limits.
- Upper and Lower Control Limits **MUST** be statistically calculated. Don't confuse them with specification limits, which are based on product requirements.
- Management, through support and action, has the opportunity to control/reduce the natural variation between the control limits.
- Make sure to select the right type of Control Chart for the right type of data (see accompanying chart).
- Data must be kept in the exact sequence as it was gathered, otherwise it is meaningless.
- Do not "tweak" the process beyond standard procedures while you are gathering the data; the data must reflect how it runs *naturally*.

> **Process Capability:** When you need to determine whether the process, given its natural variation, is capable of meeting established (customer) specifications.

PROCESS CAPABILITY

Being in control is not enough. An "in-control" process can produce bad product. True improvement of a process comes from balancing repeatability and consistency with the capability of meeting your customer requirements, otherwise known as Process Capability.

In order to objectively measure the degree to which your process is or is not meeting those requirements, capability indices have been developed to graphically portray that measure. Capability indices let you place the distribution of your process in relation to the specification limits.

FORMULAS FOR CALCULATING PROCESS CAPABILITY INDICES

C_p is a simple process capability index that relates the allowable spread of the specification limits (i.e., the difference between the upper specification limit, USL, and the lower specification limit, LSL) to the measure of the actual, or natural, variation of the process, represented by $6\hat{\sigma}$, where $\hat{\sigma}$ is the estimated process standard deviation.

$$C_p = \frac{USL - LSL}{6\hat{\sigma}}$$

If the process is in statistical control, then $\hat{\sigma}$ can be estimated from the control chart:

64

$$\hat{\sigma} = \overline{R}/d_2 \quad \text{where } \overline{R} = \text{the average of the}$$
subgroup ranges
d_2 = a tabled value based on the subgroup sample size

Factors for Estimating $\hat{\sigma}$:

n	d_2	n	d_2
2	1.128	6	2.534
3	1.693	7	2.704
4	2.059	8	2.847
5	2.326	9	2.970
		10	3.078

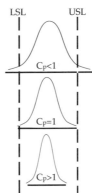

The process variation exceeds specification => defectives are being made.

The process is just meeting specification. A minimum of .3% defectives will be made, more if the process is not centered.

The process variation is less than specification, however, defectives might be made if the process is not centered on the target value.

While C_p relates the spread of the process relative to the specification width, it **DOES NOT** look at how well the process average, $\overline{\overline{X}}$, is centered to the target value. C_p is often referred to as a measure of process "potential."

The process capability indices C_{pl} and C_{pu} (for single-

65

sided specification limits) and C_{pk} (for two-sided specification limits) measure not only the process variation with respect to the allowable specification, they also take into account the location of the process average. C_{pk} is considered a measure of the process "capability" and is taken as the smaller of either C_{pl} or C_{pu}:

$$C_{pl} = \frac{\overline{\overline{X}} - LSL}{3\hat{\sigma}} \quad C_{pu} = \frac{USL - \overline{\overline{X}}}{3\hat{\sigma}} \quad C_{pk} = \min\{C_{pl}, C_{pu}\}$$

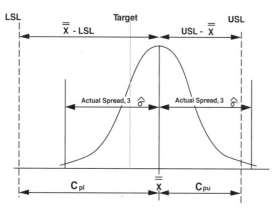

If the process is near normal and in statistical control, C_{pk} can be used to estimate the expected percent of defective material. Estimating the percentages of defective material is beyond the scope of *The Memory Jogger™* and can be found in statistical books.

Process Capability—Mfg. Example
Die Cutting Process

A Control Chart was maintained, producing the following statistics:

$\overline{\overline{X}} = 212.5$
$\overline{R} = 1.2$
$n = 5$

Spec. $= 210 \pm 3$
USL $= 213$
LSL $= 207$

$$\hat{\sigma} = \overline{R} / d_2 = 1.2/2.326 = .51$$

$$C_p = \frac{USL - LSL}{6\hat{\sigma}} = \frac{213 - 207}{6(.51)} = \frac{6}{3.06} = 1.96$$

$$C_{pl} = \frac{\overline{\overline{X}} - LSL}{3\hat{\sigma}} = \frac{212.5 - 207}{3(.51)} = \frac{5.5}{1.53} = 3.595$$

$$C_{pu} = \frac{USL - \overline{\overline{X}}}{3\hat{\sigma}} = \frac{213 - 212.5}{3(.51)} = \frac{0.5}{1.53} = .327$$

$$C_{pk} = \min\{C_{pl}, C_{pu}\} = .327$$

Since $C_{pk} < 1$, defective material is being made.

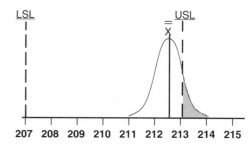

Construction/Interpretation Tips
Process Capability

- C_p index is limited to two-sided specifications.
- Some companies use the inverse of the C_p ratio

$$C_p = \frac{6\hat{\sigma}}{USL - LSL}$$

- Ask your customer which ratio is being used for interpretation.
- The use of process capability indices assumes realistic and meaningful specifications. Be sure that you and your customer have agreed upon them.
- If $C_{pl} = C_{pu}$, then the process is exactly centered!
- Many companies are establishing specific process capability targets, typically a C_{pk} of 1.33 for supplier qualification, with an expected achievement of a C_{pk} of 2.0 or higher for a long-term target.

OTHER HELPFUL TOOLS

BRAINSTORMING

All of the charting techniques are aids to thinking. They focus the attention of the user on the truly important dimensions of a problem. It is equally important, however, to expand your thinking to include **all** of the dimensions of a problem or solution. Brainstorming is used to help a group create as many ideas in as short a time as possible.

Brainstorming can be used in two ways:

1.) **Structured**—In this method, every person in a group must give an idea as their turn arises in the rotation or pass until the next *round*. It often forces even shy people to participate but can also create a certain amount of pressure to contribute.
2.) **Unstructured**—In this method, group members simply give ideas as they come to mind. It tends to create a more relaxed atmosphere but also risks domination by the most vocal members.

In both methods the general "rules of the road" are the same. The generally accepted guidelines are as follows:

- Never criticize ideas.
- Write on a flip chart or blackboard **every** idea. Having the words visible to everyone at the same time avoids misunderstandings and reminds others of new ideas.
- Everyone agrees on the question or issue being brainstormed. Write it down, too.
- Record on the flip chart in the words of the speaker; don't interpret.
- Do it quickly; 5-15 minutes works well.

68

69

NOMINAL GROUP TECHNIQUE (NGT)

When selecting which problems to work on and in what order, it often happens that the problem selected is that of the person who speaks the loudest or who has the most authority. This often creates a feeling in the team that *"their"* problem will never be worked on. This can lead to a lack of commitment to work on the problem selected, and the selection of the *"wrong"* problem in the first place. The **Nominal Group Technique** tries to provide a way to give everyone in the group an equal voice in problem selection. The steps in the process are as follows:

1. Have everyone on the team write (or say) the problem that he/she feels is most important. If members of the team do not write the problem out, you need to get them written on a flip chart or blackboard (or somewhere visible), as they are being communicated. If people do produce written problems, collect them when they are finished. Everyone may not feel comfortable writing, but it may make them feel safer talking about sensitive problems at the beginning.
2. Write the problem statements where the team can see them.
3. Check with the team to make sure that the same problem hasn't been written twice (but may be in slightly different words). If a problem is repeated combine them into one item.
4. Ask the team members to write on a piece of paper the letters corresponding to the number of problem statements the team produced. For example, if you ended up with five problem statements, everyone would write the letters "A" through "E" on the paper.
5. Make sure that each problem statement has a letter in front of it. Ask the team members to vote on which problem is most important, then write "5" next to the letter. For example, the problem list may look like this:
   A. Space
   B. Safety
   C. Housekeeping
   D. Quality going down
   E. No preventive maintenance

Each member's paper would look like this:
   A. _____
   B. _____
   C. _____
   D. _____
   E. _____

So, if someone thought *"quality is going down"* was the #1 problem, it would look like this:
   A. _____
   B. _____
   C. _____
   D. ___5___
   E. _____

Everyone then has to complete the list by voting what's second most important, third most important, etc.
   A. _2, 5, 2, 4, 1_
   B. _1, 4, 5, 5, 5_
   C. _4, 1, 3, 3, 4_
   D. _5, 2, 1, 1, 2_
   E. _3, 3, 4, 2, 3_

An alternative ranking approach involves the "one half plus one" rule. Especially when dealing with many items, it may be necessary to limit the items to be considered. This rule suggests ranking only one half of the items plus one. For example, if 20 items were generated, team members would rate only 11 ideas.

6. Add up each line of numbers across. The item with the **highest** number is the most important one to the total team. In this case "B" (Safety) would be the most important item with a total of "20." You would add up the numbers for each item and put them in order.

7. You would then work on item B first, and then move through the list.

FORCE FIELD ANALYSIS

How does change occur, either personally or organizationally? It's a dynamic process. It suggests movement, either from *"time A"* to *"time B"* or *"condition x"* to *"condition y."* etc. Where does the energy for this *"movement"* come from? One approach is to view change as the result of a struggle between forces that are seeking to upset the status quo. This view is taken in the work of Kurt Lewin, who developed a technique called *"Force Field Analysis."* In it, Lewin proposed that *"driving forces"* move a situation toward change while *"restraining forces"* block that movement. When there is no change, the opposing forces are equal or the restraining forces are too strong to allow movement.

Consider the practical example of *"losing weight"*:

Driving Forces	Restraining Forces
Health threat →	← Lack of time
Cultural obsession with being thin →	← Genetic traits
Plenty of thin role models →	← Unsympathetic friends & family
Embarassment →	Lack of money for
Negative self-image →	← exercise
Positive attitude toward exercise →	← Lack of interest
Lack of temptation →	← Bad advice
Clothes don't fit →	Years of bad eating ← habits
	Amount of sugar in ← any prepared food

If the restraining forces are stronger than the driving forces, then the desired change will not happen. It stands to reason that some change (lost weight) will occur if the driving forces are more powerful than those on the restraining side of the ledger.

Why does Force Field Analysis help make change happen?

1.) It forces people to think together about all the facets of a desired change; it encourages creative thinking.

2.) It encourages people to agree about the relative priority of factors on each side of the "balance sheet." The team can use the Nominal Group Technique to reach consensus quickly.

3.) It provides a starting point for action.

How does Force Field Analysis accomplish this last task? One can approach change either from the perspective of strengthening *"driving forces"* or reducing the *"restraining forces."* Strengthening the positive often has the unexpected result of reinforcing the negative. Have you ever seen a situation where someone is told repeatedly that *"X, Y, or Z"* is bad for him or her? Instead of the desired improvement, it often strengthens resistance. It has been shown that the most effective tactic is to diminish or eliminate a restraining force. In our example, it would be much more helpful to deal with *"lack of time"* than to constantly remind someone that his or her *"clothes don't fit"* like they used to.

If done honestly, Force Field Analysis can be a helpful aid to thinking and a strategic tool for change.

MANAGEMENT PRESENTATION CHECKLIST

— Decide who the key decision makers are.

— Invite the key decision makers to the presentation.
  • By memo if appropriate
  • By phone the day before to get confirmation

— Arrange for the room.
  • Right size; large enough to be comfortable; not so large that group members are lost
  • Right atmosphere; the ideal is bright, quiet, clean, and informal

— Arrange for audiovisual (AV) equipment.
  • Flip chart
  • Overhead projector
  • Screen
  • Markers

— Decide which charts will help the presentation.

— Prepare charts and AV materials.
  • Assign responsibility
  • Keep AV materials few in number and simple in design

— Assign presentation responsibilities.

— At the presentation:
  • Speak clearly and slowly
  • Listen to questions carefully; you will get an idea of what to stress in the presentation
  • Summarize recommendations simply and clearly

74

PIE CHART

Pie Charts are simply graphs in which the entire circle represents 100% (not 360 degrees) of the data to be displayed. The circle (pie) is divided into percentage *slices* that clearly show the largest shares of data. This is useful in the same way as a **Pareto Chart**. The Pie Chart is sometimes even more useful since it is widely used to display data on TV or in the newspapers. As with all other graphs, be sure to clearly mark the subject matter, dates if needed, the percentages within the *slices*, and what each slice represents.

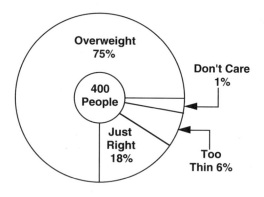

Pie Chart
Survey of Feelings
On Weight

75

STRATIFICATION*

A technique called Stratification is often very useful in analyzing data to find improvement opportunities. Stratification helps analyze cases in which data actually masks the real facts. This often happens when the recorded data is from many sources but is treated as one number.

For example, data on minor injuries for a plant may be recorded as a single figure either rising or falling. But that number is actually the sum total of injuries:

- By Type: cuts and burns
- By Location: eyes, hands, feet, etc.
- By Department: maintenance, shipping, production, etc.

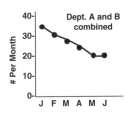 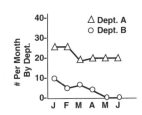

Stratification breaks down single numbers into meaningful categories or classifications to focus corrective action.

*See "Different Uses Of A Pareto Chart," pages 20-22, for stratification in pareto analysis. Also see pages 172-174 of the Western Electric Statistical Quality Control Handbook for examples of stratification in control chart patterns.

76

ADDITIONAL BAR GRAPHS
Compound Bar Graph

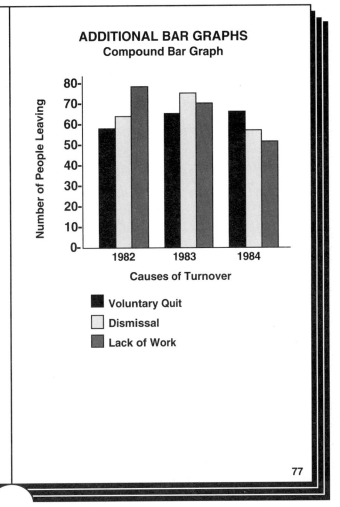

77

301

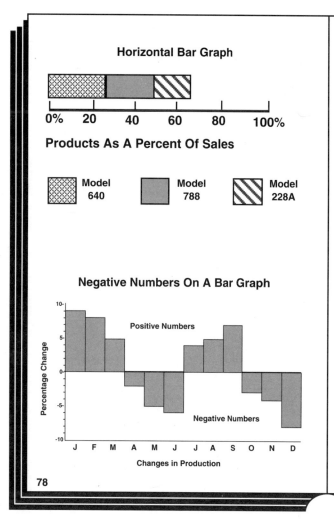

Horizontal Bar Graph

Products As A Percent Of Sales

Model 640   Model 788   Model 228A

Negative Numbers On A Bar Graph

Positive Numbers

Negative Numbers

Percentage Change

J F M A M J J A S O N D

Changes in Production

78

Construction/Interpretation Tips
Data Points To Ponder

- The aim of data-based problem solving is not to gather **more** data but **meaningful** data.
- The collection and proper use of data short-circuits much of the interpersonal conflict that happens in groups.
- Data can be used to:
  — Understand the current **actual** situation (good and bad)
  — Regulate and modify the process
  — Accept or reject a product or process
- Bad data is worse than no data.
- Data should ideally be based on a truly random sample in which each event or piece has an equal chance of being observed or selected.
- Data that is compared must be gathered **consistently**.
- There are two major types of data, **measured/continuous** and **counted/discrete**.

Measured/Continuous	Data that's measured on a continuous scale such as length, weight, time, or temperature.
Counted/Discrete	Data that is an accumulation of observations of a particular characteristic such as the number of defects, units sold, or typing errors.

- Every data collection document should include:
  — Name of person(s) collecting data
  — Date of collection
  — Time span covered, e.g., time of day, shift
  — Location of data collection, e.g., department, office
  — Instruments (if applicable) or any other methods used

79

Construction/Interpretation Tips Common Mistakes To Avoid

- Don't bias the results by the method of sampling. Try to gather sample data in as random a way as possible, e.g., don't take all the pieces from the top of the box.
- Don't confuse control limits with specification limits. Control limits are statistically determined while specification limits are based on what is needed or desired.
- Don't make it more complicated than it needs to be. Use the simplest appropriate tool.
- Don't collect too much or too little data. Don't collect data for a week when a day will do or vice versa.
- Don't overcomplicate graphs. Keep them simple and clear so that the message is apparent to the viewer.
- Don't confuse samples with populations.
- Don't blindly interpret graphs the same way in different situations. Use common sense, e.g., frequency of occurrence may not always be the most significant measure under the Pareto principle.
- Don't stand on a single statistic. Have other supporting evidence, e.g., find the range, not just the mean.
- Don't hesitate to seek help when a situation seems too complex or confusing for you to handle. Many companies today have professional statisticians who can help you collect and analyze information in the most efficient and effective way possible.

GLOSSARY OF TERMS USED IN STATISTICAL PROCESS CONTROL

*ATTRIBUTES are qualitative data that can be counted for recording and analysis. Examples include characteristics such as the presence of a required label and the installation of all required fasteners. Other examples may include characteristics that are inherently measurable (i.e., could be treated as variables). Where the results are recorded in a simple yes/no fashion, (such as acceptability of a shaft diameter when measured on a go/no-go gage), p, np, c, and u charts are used rather than an X & R chart.

AVERAGE or mean is the most common expression of the centering of a distribution. It is signified by $\overline{X}$ and is calculated by totaling the observed values and dividing by the number of observations

$$\overline{X} = \frac{(X_1 + X_2 + ...+ X_n)}{n}$$

BIMODAL DISTRIBUTION is one which has two identifiable curves within it, indicating a mixing of two populations such as different shifts, machines, workers, etc.

*COMMON CAUSE is a source of variation that is always present; part of the random variation inherent in the process itself. Its origin can usually be traced to an element of the system which only management can correct.

*CONTROL CHART is a graphic representation of a characteristic of a process, showing plotted values of some statistic gathered from that characteristic, and one or two control limits. It has two basic uses: as a judgment to determine if a process was in control, and as an aid in achieving and maintaining statistical control.

80

81

*CONTROL LIMIT is a line (or lines) on a control chart used as a basis for judging the significance of the variation from subgroup to subgroup. Variation beyond a control limit is evidence that special causes are affecting the process. Control limits are calculated from process data and are not to be confused with engineering specifications.

*DETECTION or inspection is a past-oriented strategy that attempts to identify unacceptable output after it has been produced and separate it from the good output. (See also PREVENTION)

DISTRIBUTION is the population (universe) from which observations are drawn, categorized into cells, and form identifiable patterns. It is based on the concept of variation that states that anything measured repeatedly will arrive at different results. These results will fall into statistically predictable patterns. A *bell-shaped curve* (normal distribution) is an example of a distribution in which the greatest number of observations fall in the center with fewer and fewer observations falling evenly on either side of the average.

FORCE FIELD ANALYSIS is a technique developed by Kurt Lewin that displays the Driving (Positive) and Restraining (Negative) forces surrounding any change. This is displayed in a "balance sheet" format.

FREQUENCY DISTRIBUTION is a statistical table that presents a large volume of data in such a way that the central tendency (average/mean/median) and distribution are clearly displayed.

NOMINAL GROUP TECHNIQUE is a weighted ranking technique that allows a team to prioritize a large number of issues without creating "winners" and "losers."

*NONCONFORMITIES are specific occurrences of a condition that does not conform to specifications or other inspection standards; sometimes called discrepancies or defects. An individual nonconforming unit can have the potential for more than one nonconformity (e.g., a door could have several dents and dings; a functional check of a carburetor could reveal any of a number of discrepancies). The c and u charts are used to analyze systems producing nonconformities.

POPULATION is the universe of data under investigation from which a sample will be taken.

*PREVENTION is a future-oriented strategy that improves quality by directing analysis and action toward correcting the production process. Prevention is consistent with a philosophy of never-ending improvement.

*PROCESS is the combination of people, machine and equipment, raw materials, methods, and environment that produces a given product or service.

PROCESS CAPABILITY is the measured, built-in reproducibility (consistency) of the product turned out by the process. Such a determination is made using statistical methods, not wishful thinking. The statistically determined pattern or distribution can only then be compared to specification limits to decide if a process can consistently deliver product within those parameters.

RANGE is a measure of the variation in a set of data. It is calculated by subtracting the lowest value in the data set from the highest value in that same set.

RUNS are the patterns in a Run Chart or Control Chart within which a number of points line up on only one side of the central line. Beyond a certain number of consecutive points (statistically based) the pattern becomes unnatural and worthy of attention.

*SAMPLE is one or more individual events or measurements selected from the output of a process for purposes of identifying characteristics and performance of the whole.

*SIGMA $\hat{\sigma}$ is the Greek letter used to designate the estimated standard deviation.

*SPECIAL CAUSE is a source of variation that is intermittent, unpredictable, unstable; sometimes called an assignable cause. It is signalled by a point beyond the control limits.

*SPECIFICATION is the engineering requirement for judging acceptability of a particular characteristic. Chosen with respect to functional or customer requirements for the product, a specification may or may not be consistent with the demonstrated capability of the process (if it is not, out-of-specification parts are certain to be made). A specification should never be confused with a control limit.

*STANDARD DEVIATION is a measure of the spread of the process output or the spread of a sampling statistic from the process (e.g., of subgroup averages), denoted by the Greek letter $\hat{\sigma}$ (sigma) for the estimated standard deviation.

*STATISTICAL CONTROL is the condition describing a process from which all special causes have been removed, evidenced on a control chart by the absence of points beyond the control limits and by the absence of non-random patterns or trends within the control limits.

*STATISTICAL PROCESS CONTROL is the use of statistical techniques such as Control Charts to analyze a process or its output so as to take appropriate actions to achieve and maintain a state of statistical control and to improve the capability of the process.

STRATIFICATION is the process of classifying data into subgroups based on characteristics or categories.

TRENDS are the patterns in a Run Chart or Control Chart that feature the continued rise or fall of a series of points. Like Runs, attention should be paid to such patterns when they exceed a predetermined number (statistically based).

*VARIABLES are those characteristics of a part which can be measured. Examples are length in millimeters, resistance in ohms, closing effort of a door in kilograms, and the torque of a nut in foot pounds. (See also ATTRIBUTES)

*VARIATION is the inevitable difference among individual outputs of a process. The sources of variation can be grouped into two major classes: Common Causes and Special Causes.

*From Ford Motor Company's Q101 (1983). This is a handbook of quality requirements for its manufacturing plants and outside vendors.

84

85

The Memory Jogger™ 9000

The *Memory Jogger™ 9000* is the best source for you and everyone in your organization to learn how to comply with the ISO 9000 Standard and QS-9000 Requirements. It will show you how to: participate in identifying the work that you do that affects the quality of your organization's products and services; improve your work processes; write procedures that describe your work; and much more.

Code: 1060E   Price: $7.95

The Team Memory Jogger™

Easy to read and written from the team member's point of view, *The Team Memory Jogger™* goes far beyond basic theories to provide you with practical nuts-and-bolts action steps on preparing to be an effective team member, how to get a good start, get work done in teams, and when and how to end a project. *The Team Memory Jogger™* also teaches you how to deal with problems that can arise within a team. It's perfect for all employees at all levels.

Code: 1050E   Price: $7.95
Quantity discounts are available.

Coach's Guide to The Memory Jogger™ II CD-ROM Package

The *Coach's Guide Package* makes it easier than ever to use *The Memory Jogger™ II* as a key resource in your effective training efforts. You can get your teams to better use the basic quality control tools and management and planning tools so that they can achieve their objectives, and learn to rely less on the facilitators and more on their own self-sufficiency in solving problems.

The tools included are: Activity Network, Affinity, Brainstorming, Cause & Effect, Check Sheet, Control Chart, Flowchart, Force Field Analysis, Histogram, Interrelationship Digraph, Matrix, Nominal Group Technique, Pareto, Prioritization Matrices, Process Capability, Radar, Run, Scatter, and Tree.

Each tool has a set of overheads that includes these features: • a summary of the steps for using the tool • an overview of the steps in flowchart form • illustrations of the tool at different steps • finished examples of the tools

The CD-ROM Package includes:
- 1 *Coach's Guide*
- 1 CD-ROM disk with 187 overheads in Microsoft PowerPoint™
- 5 copies of *The Memory Jogger™ II*

You can print the overheads as you need them and you can customize them too!

Code: 1046   Price: $185
Quantity discounts are available.

Quantity discounts are available. Please call for details!
800-643-4316 or 978-685-6370 or Fax: 978-685-6151
E-mail: service@goal.com or Web site: http://www.goalqpc.com